SCOPES II
THE GREAT DEBATE

CREATION vs. EVOLUTION

*S*COPES II
the Great Debate

Bill Keith

HUNTINGTON HOUSE, INC.
1200 N. MARKET STREET, SUITE G
SHREVEPORT, LOUISIANA 71107
(318) 222-1350

First printing 1982
Second printing 1983
Third printing 1984
Fourth printing 1985

The material in chapter 9 — pages 124-140 — is taken from copyrighted material copyrighted by Mott Media, 1983, and also appears in the book *The Creator in the Courtroom* by Dr. Norman Geisler.

Printed in the United States of America

ISBN: 0-910311-01-3

Library of Congress Catalog Card Number: 82-082837

To my wife, Lowayne,
who stands close by my side
in this great battle for academic freedom.

degree majoring in law, but rather than go through a legal apprenticeship, he answered a newspaper ad for a teacher in Dayton.

Some historians believe he was recruited by Rep.

Prologue

On a hot July day in 1925, William Jennings Bryan and Clarence Darrow fought one of the great legal battles of history.

It took place in Dayton, Tennessee, and some called it the "Trial of the Century." Others referred to it as the "monkey" or "Scopes" trial. The issue was whether Darwinian evolution could be taught in the public schools of Tennessee.

The legal scenario featured three now-famous characters. They were:

• John T. Scopes, a 24-year-old biology teacher on trial for teaching evolution which was prohibited by state law.

• William Jennings Bryan, the silver-tongued orator and three-time presidential candidate who argued for the state.

• Clarence Darrow, the Chicago, Illinois, lawyer who defended Scopes.

Scopes was born in Kentucky where he earned a bachelor's degree majoring in law. But rather than go through a legal apprenticeship, he answered a newspaper ad for a teacher in Dayton.

Some historians believe he was recruited by Roger Baldwin, the founding father of the American Civil Liberties Union, just to test the Tennessee law governing evolution. Baldwin did hand-pick Darrow to defend Scopes.

Darrow, who performed with great skill, said during the

trial that teaching only one theory of origins is sheer bigotry. He also asked the question: "Can the human mind be limited by law in its inquiry after truth?" His question parallels the question creation scientists are asking today.

Bryan — known throughout America as "The Great Commoner" — was one of history's great orators. During the 1896 Democratic National Convention he delivered his "Cross of Gold" speech which denounced the gold standard of the time and catapulted him into the presidential nomination at the age of 36.

Both Scopes and Bryan had close ties to Louisiana where the battle called Scopes II soon will be fought. The Great Commoner delivered two anti-evolution addresses in Shreveport in 1900 and 1924. Thousands of people crowded around the courthouse and into the fairgrounds to hear him speak. Scopes moved to Shreveport in 1940 and lived there until his death in 1970. The Tennessee Legislature had repealed the law forbidding the teaching of evolution three years earlier.

The three leading characters in Scopes I were enshrined in the minds and hearts of the American people through the book *Inherit the Wind,* based on the famous trial. It was also made into a movie.

Darrow, though he lost the case, became a celebrated trial lawyer. Scopes gave up the teaching profession and became a successful geologist. Bryan died a few days after the trial concluded — in Dayton, Tennessee.

The new drama — similar to the famous Scopes Trial — will take place in Louisiana. Some have called it "Scopes II" or "Scopes I in Reverse."

The characters have changed. There will be no Bryans, Darrows or Scopes in the Louisiana trial.

Although Scopes II has a different array of characters, the issue of academic freedom is the same.

The question in Dayton, Tennessee, in 1925 was: Should the scientific evidences for evolution be taught to public school children along with those evidences for creation?

The question today in Louisiana — and throughout America — is: Should the scientific evidences for creation be given equal time with those evidences for evolution?

Scopes II, the great debate, will take a step toward answering that question.

Contents

also adopted a program that would give equal time to
creation-science whenever and wherever evolutionary
dogma is taught in the schools.

CHAPTER 1

Victory in Louisiana

The great debate on the subject of creation vs. evolution is one of the burning issues of our times. Nothing in the academic world is more important for it affects every child who enters a public school.

Most children grow up in homes where they are taught that the earth and everything in it was created by a Supreme Being. Yet when they enter public schools, suddenly they are confronted with evolution, a concept quite foreign to them.

This experience does damage to the young mind as the evolutionary indoctrination begins in the first grade. The damage continues as the student is bombarded with evolutionary propaganda all the way through the elementary and secondary schools of our land.

We tried to do something about that problem in Louisiana.

Let me tell you how our State Legislature overwhelmingly adopted a program that would give equal time to creation-science whenever and wherever evolutionary dogma is taught in the schools.

A hush settled over the Louisiana Senate chambers. Debate ended. Legal maneuvering ceased. People in the galleries waited.

Senate President Michael O'Keefe called for a final vote

on Senate Bill 86. It would mandate the balanced treatment of creation-science and evolution-science in the state's public school classrooms.

It had been a long afternoon. Opponents had used every maneuver known to the legislature to kill the bill. One senator made a motion that it be returned to the calendar. The motion failed. Another asked that it be sent back to the Education Committee for further study. His motion also failed.

The majority of the senators repeatedly rejected amendments that would have crippled the bill and destroyed its effectiveness.

Suddenly the heated debate and oratorical wrangling ended. It was time to vote.

"Those who favor the bill will vote yea and those who oppose will vote nay and the secretary will open the machines," the president said.

Everyone present looked intently at the giant board at the upper west end of the chamber. It would record each senator's vote.

The senators pressed the voting buttons on their desks. Both green and red lights appeared on the board.

"The bill carries 26 to 12," the president proclaimed, as sighs of relief reverberated through the chambers. He then instructed the Senate secretary to forward the bill to Governor Dave Treen for his signature.

Thus, on July 7, 1981, at about 4 p.m., Senate Bill 86 cleared the final legislative hurdle. It was the most hotly contested bill of the legislative session. Creationists had won a great victory.

The victory culminated 12 months of intensive debate in various committee hearings between creationists and evolutionists.

Scribes and pundits had predicted the bill would never pass. Critics laughed, even scoffed at it. Newspaper editorials almost universally opposed it as did many members of the scientific community and the American Civil Liberties Union.

Yet the people of the state sent a message to the legislators at the capitol in Baton Rouge. Tens of thousands of Louisianians wrote letters, made phone calls and sent telegrams urging passage of the bill. The legislators heard — and listened.

Passage of the Balanced Treatment Act also culminated my own three-year odyssey in search of scientific truth as it relates to academic freedom.

That odyssey began in 1978 before I was elected to the Senate. It was while I was still the city editor of the afternoon newspaper in Shreveport, Louisiana.

One evening when I arrived home from work, my wife, Lowayne, told me a very bizarre story. It concerned our twelve-year-old son Richard and his science teacher.

My wife told me his science class had been studying the subject of origins. On that particular day the teacher directed her very first question to our son, asking him how the world was formed. His answer reflected the opinion of most twelve-year-old boys or girls when he said, "It was created."

Somewhat startled, the teacher asked him if he had read his textbook assignment.

"Yes," he replied.

"What did your textbook say?" she asked.

"The textbook said we came from monkeys," he answered. "But I don't believe it — I believe we were created."

The teacher could have just smiled and said something like, "Well, there are different ideas about how the world came into being."

But apparently she was very threatened, even intimidated by my son's answers. She began to ridicule and harass him in front of his peers. She ordered him never to say that again in her classroom or she would take him to the principal for discipline.

My wife was in tears as she concluded the story. It shocked and angered me. At that time I thought we still had academic freedom in this land. That the public schools were a marketplace for ideas. Apparently that was no longer true.

That night I told my wife: "I have no plans to run for political office but if I ever do, and win, I'm going to do something about school children being abused because of their beliefs."

Later I learned the science teacher was a substitute and there for only one day so I did not speak to her about the unfortunate incident. But I felt that if she wanted to debate the subject of origins she should debate me and not my twelve-year-old son.

Ironically, I was still an evolutionist when our son's incident with the teacher took place. Oh, I believed in a Creator-controlled evolution. Why not? My teachers in high school told me it was true and my college professors confirmed it.

Yet several questions troubled me about what had happened to my son.

Should my son be required to be an evolutionist in order to get along with his teacher? Should he be encouraged to believe in evolution just because his father and his teacher believed it was a rational explanation of origins? Did he have the right to choose for himself? How could he choose for himself if only one concept was presented to him?

That one incident in a science classroom in a public school revolutionized my life. I began to study the subject of origins. My search opened up a new world of understanding, not only about origins but also about academic freedom and scientific integrity.

After only a few months I became convinced that Darwinian evolution is a hoax, perhaps the greatest hoax of the twentieth century. A fairy tale for adults comparable to the story of the frog that turned into a prince.

My search for the truth convinced me that there is not

one shred of proof supporting evolution. Rather, it is a metaphysical research program. Religion rather than science. Voodoo logic rather than sound reason. Not science but science fiction.

Evolution is really a cute little way to make monkeys out of men and that is just what has happened to most of us.

Here is my own version of the fairy tale:

"Once upon a time in a far, far away land some dead elements met. That accidental meeting among the dead created life — a living molecule. But it wasn't just any average, everyday molecule. It could reproduce itself. As a matter of fact, it evolved into even more complex organisms and they became fish. The fish became amphibia which became reptiles which became mammals. The mammals became monkeys and men."

If you believe that story then let me tell you about the Easter Bunny, Santa Claus and the Tooth Fairy.

Probing deeper into the subject of academic freedom regarding origins, I made some remarkable discoveries:

• There are bonafide scientific evidences, completely separate from the Genesis account in the Bible, which point to creation.

• Although evolution is no more than a theory, many science teachers present it as fact to school children.

• All scientific data pointing to creation is summarily censored out of the school science curriculum.

• Evolution is a cardinal tenet of the religion of secular humanism.

• There are only two basic viewpoints on the subject of origins — creation and evolution — and there is no third scientific view.

• Many of the greatest scientists of history were creation-scientists.

• It is an abridgment of academic freedom and therefore patently unconstitutional to teach only one view of origins.

• Many well-known scientists today are abandoning

their belief in evolution.

• The mathematical odds against the universe creating itself are 40,000 trillion, trillion, trillion, to one. It would require several thousand pages of paper just to print these odds.

• The Darwinian evolution most of us learned in high school is no longer considered valid by today's leading spokesmen on the subject of evolution. They now couch their metaphysical belief in evolution in such terms as "punctuated equilibria." This means that the various species remained virtually the same for millions of years, then suddenly took "quantum leaps." Or, just overnight, an ape-like creature, through a "quantum leap," became a man. This theory requires no "missing link" or transitional form.

• The methods for dating the age of the earth are based on presuppositions which may not be valid.

• Relatively new scientific studies reveal the earth is far younger than the billions of years necessary for evolution to have taken place.

• Since Charles Darwin published his famous book *The Origin of Species* in 1859, not one "missing link" or half-man and half-monkey-like creature has ever been found. Numerous skeletons found by paleontologists have been reported to be "missing links" but all have been proven to be frauds or mistakes. However, pictures of them have appeared in public school textbooks as proof of evolution.

• If evolution were true and it took tens of millions of years for man to have evolved from an ape-like creature, then there should be tens of millions of half-man and half-monkey-like creatures in the fossil records. Yet there are none.

• The Declaration of Independence says: "We hold these truths to be self-evident, that all men are *created* equal, that they are endowed by the Creator with certain unalienable rights, that among these are Life, Liberty and the Pursuit of Happiness."

Some very wise and courageous men signed that document on July 4, 1776. Even our Declaration of Independence speaks of creation and a Creator. It doesn't say: "All men evolved from monkey-like creatures."

Did you know all that? Neither did I and neither do our school children.

Science writers who produce our school textbooks present the young readers only the evolutionary point of view. Our public, tax-supported universities indoctrinate future science teachers only in evolutionary dogma.

That's censorship! And you and I pay for it with our tax dollars.

Evolutionists have a monopolistic control over the textbooks, the teacher training institutions and the classrooms. They continue to teach our children they came from apelike creatures and it is a big lie.

Now let's take a quick look at the law itself. You can decide if it is fair and valid.

The introduction to the law states that the schools of Louisiana will give balanced treatment to creation-science and evolution-science. The purpose of the law is to bar discrimination on the basis of creationist or evolutionist belief.

The law also provides a number of definitions:

• "Balanced treatment" means providing whatever information and instruction in both creation and evolution models the classroom teacher determines is necessary. It provides insight into both theories in view of the textbooks and other instructional materials available for use in the classroom.

• "Creation-science" means the scientific evidences for creation and inferences from those scientific evidences.

• "Evolution-science" means the scientific evidences for evolution and inferences from those scientific evidences.

• Balanced treatment of these two models shall be given in classroom lectures taken as a whole for each course, and also in textbook and library materials in the areas of science, the humanities and other educational programs. It is limited to those courses which actually deal with the subjects of the origin of man, life, and the earth, or the universe.

• When creation or evolution is taught, each shall be taught as a theory, rather than as proven scientific fact.

• Public schools within Louisiana and their personnel shall not discriminate by reducing a grade of a student or by singling out and publicly criticizing any student who demonstrates a satisfactory understanding of both evolution-science or creation-science and who accepts or rejects either model in whole or part.

• No teacher in public elementary or secondary school or instructor in any state-supported university in Louisiana, who chooses to be a creation-scientist or to teach scientific data which points to creationism shall, for that reason, be discriminated against in any way by any school board, college board, or administrator.

• The law does not require any instruction in the subject of origins. It simply permits instruction in both scientific models if the public schools choose to teach either.

• Each city and parish school board shall develop and provide to each public school classroom teacher in the system a curriculum guide on the presentation of creation-science.

• The governor shall designate seven creation-scientists who shall provide resource services in the development of curriculum guides to any city or parish (county) school board upon request. Each such creation-scientist shall be designated from among the full-time faculty members teaching in any state college or university in Louisiana.

These were the basic tenets of the Balanced Treatment Act. It was a giant step forward for academic freedom, but came under strong attack by the evolutionary forces in the state.

Recently I received a letter from a friend in California who wrote: "One of my young friends was practically in tears when she described how her public school teacher in Northern California asked all the students who believed in creation to raise their hands.

"The teacher then proceeded to 'sneer at and ridicule' those children who lifted their hands."

That kind of indoctrination is so unfortunate and must be stopped!

We are trying to get it stopped in Louisiana.

CHAPTER 2

The Great Debate

Our first attempt in 1980 to pass a law providing equal time for creation-science along with evolution was a fiasco.

But we learned a great lesson — the legislators and the people of the state had to be educated on the subject. We also discovered that the evolutionists are formidable opponents and not to be taken lightly.

Several recent events had helped our cause:

• Presidential candidate Ronald Reagan just previously had told a group meeting in Dallas he believed evolution should be taught as a theory in the schools. He also endorsed teaching creation-science. His endorsement was widely publicized in Louisiana and across America.

• When certain newspapers in the state first learned about the bill they began taking polls on the subject. Through those polls we learned that the people of the state approved, by a large margin, balanced treatment. Louisiana has about four million people. The polls estimated that three million of them were with us.

• One poll conducted in my own senatorial district even astounded me. More than 80 percent of the people were with me on the subject of balanced treatment. Some 53 percent said they would like to see evolution removed from the school curricula in our area.

Polls throughout America consistently indicated that 75 percent of the people favor equal time for creation-science.

Those statistics irritated the evolutionists — and even some editorial writers at newspapers. But we were greatly encouraged.

Our first confrontation with the evolutionists occurred on June 19, 1980. Hundreds of creationists packed the Senate Education Committee room and the overflow stood outside to try to hear the proceedings. The senators on the committee were impressed with the large numbers of people who were interested in the matter.

I had invited Luther Sunderland to testify before the committee on behalf of creation-science and to function as our major speaker. He is an aerospace engineer with the General Electric Company in New York, and well known expert on creation vs. evolution.

Sunderland made one of the most dramatic presentations ever witnessed at the Louisiana capitol.

The committee chairman called for the debate to begin on the bill and asked me to begin our presentation.

After making a few introductory remarks, I introduced Sunderland to the committee members.

His presentation dealt primarily with how both concepts of origins — creation and evolution — could be taught in the public schools.

He then proceeded with color-slide documentation on the subject matter. That same presentation had been given to the New York State Board of Regents and was featured by Walter Cronkite on the CBS Evening News.

"Since the beginning of time, man has been very concerned about origins and he has been able to think about two concepts or two general frameworks for how everything came about," Sunderland said. "One of these is creation and the other is evolution. Now someone might come up with a third idea, but these are the only two basic concepts we have ever heard of.

"It is true that neither of these is truly a scientific theory because neither meets any of the qualifications of a true scientific theory. Neither evolution nor creation can be repeated. They cannot be observed. They cannot be tested. . . . You cannot rebuke evolution. You cannot test or observe it. It takes too many thousands of years to bring out any significant change.

"If creation were true, it is a unique historic event not witnessed by any human observers. Therefore, neither of these are truly scientific theories. They are, however, proper scientific type hypotheses or models and we can take a look at the body of scientific data in the world and compare it against these two general models — creation and evolution — to see which fits the facts better.

"We need to define for the students these two general models. This is really about all that is involved in teaching the two.

"The evolution model is taught now in virtually every public school system from kindergarten through university. This is the concept that all living things have originated from a single source or single cell which originated due to property inherent in matter alone. This first cell spontaneously arose from non-living matter. The universe likewise spontaneously originated out of nothing or perhaps some inert matter . . . over a long period of time.

"There is no confusion on the issue. This is called the particles-to-people, the molecule-to-man concept."

While Sunderland was speaking he used the visual effects to illustrate his various points.

He simply pointed out that the creation model of origins is acceptable for teaching in the public schools.

Then he really got the committee's rapt attention when he said that every basic law in science indicates the universe could not have created itself.

"Therefore, this model postulates that everything was created by some intelligence or power external to the universe," Sunderland explained. "The original organisms that were created . . . were functionally complete when created — including man.

"Man was ready to go when he hit this earth, so was every other originally created organism. These organisms had within their genes a capability to vary and survive in a changing environment, whether it was hot or cold, wet or dry. They would roll with the punches and survive."

Sunderland explained what he meant by the statement. He said that the fossil record reveals that when man appears he is complete man, horses complete horses and dogs complete dogs. He added that evolution does take place within a type or taxon but never outside a type or taxon and that is decidedly proven in the fossils.

According to Sunderland, the key to what really happened in the past is in the fossil record. He explained that two-thirds of the earth's surface is covered by sedimentary rocks which were formed when sediments and sand settled under water. They were later hardened by some agent, trapping within them billions and billions of fossils.

"We can evaluate the fossil record and other scientific evidences against the creation model and the evolutionary model," he explained to the committee. "That is all we do in the science classroom. There is no religion brought into it whatsoever . . . just these two basic concepts that can be derived strictly from observing the scientific data."

Sunderland pointed out to the committee that if evolution were true, we should be able to find a progression of single-celled, two-celled, or three-celled organisms becoming fish; the fish growing feet and legs thus becoming amphibians; the amphibians developing hard shells and eggs, then becoming reptiles; the reptiles growing feathers then becoming birds, or growing hair and becoming mammals. That, he said, is what we would expect to find in the fossils to prove evolutionary process.

"If creation were true, you would expect to find every organism fully developed and fully functional when it first appeared in the fossil record," he said. "A dragonfly should be 100 percent dragonfly; a fish, a fish; a man, a man."

According to Sunderland, it is necessary to refer to the

geologic column to see exactly what happened. He jokingly remarked that the age of the earth has been doubling on the average of once every 15 years due to the changing estimates made by the evolutionists.

Sunderland explained to the committee that the very bottom or oldest rocks in the fossil record are called Cambrian. They are the deepest rocks that contain any fossils and are very complex.

He then read a letter he had received from Dr. Colin Patterson, a paleontologist with the British Museum of Natural History in London. Dr. Patterson had written a book which was considered a monumental work. Sunderland read the book but noticed that no "missing links" or transitional forms were produced in the work, so he wrote Dr. Patterson to ask why.

Following is the reply from Dr. Patterson — an evolutionist and one of the great scholars in England — which Sunderland presented to the committee.

"Thanks for your letter of 5th March, and your kind words about the Museum and my book. I held off answering you for a couple of weeks, in case the artwork you mention in your letter should turn up, but it hasn't.

"I fully agree with your comments on the lack of direct illustration of evolutionary transitions in my book. If I knew of any, fossil or living, I would certainly have included them. You suggest that an artist should be asked to visualize such transformations, but where would he get the information from? I could not, honestly, provide it, and if I were to leave it to artistic license, would that not mislead the reader?

"I wrote the text of my book four years ago. If I were to write it now, I think the book would be rather different. Gradualism is a concept I believe in, not just because of Darwin's authority, but because my understanding of genetics seems to demand it. Yet Gould and the American Museum people are hard to contradict when they say that there are no transitional fossils. As a paleontologist myself,

I am much occupied with the philosophical problems of identifying ancestral forms in the fossil record. You say that I should at least 'show a photo of the fossil from which each type organism was derived.' I will lay it on the line — there is not one such fossil for which one could make a watertight argument. The reason is that statements about ancestry and descent are not applicable in the fossil record. Is *archaeopteryx* the ancestor of all birds? Perhaps yes, perhaps no; there is no way of answering the question. It is easy enough to make up stories of how one form gave rise to another, and to find reason why the stages should be favoured by natural selection. But such stories are not part of science, for there is no way of putting them to the test.

"So, much as I should like to oblige you by jumping to the defense of gradualism, and fleshing out the transitions between the major types of animals and plants, I find myself a bit short of the intellectual justification necessary for the job. . . ."

Sunderland agrees with Patterson that there is no evidence of gradual progression in the Cambrian rocks but there is evidence of sudden appearances of fully formed fossils.

"Now notice, I have not mentioned religion," Sunderland said, adding that religion need not be discussed when the two models — creation and evolution — are presented in the schools.

"Objective science is simply the evaluation of scientific data letting the students make up their own minds," he said. "That is all that we are advocating. It works like magic in a school system. Really there are no big problems."

Sunderland also said scientifically controlled tests show the students exposed to the two-model approach learn much better than those students who learn only about evolution from textbooks where all scientific evidences for creation have been censored.

"All we are asking for is open, unbiased education in the

classroom," he said. "We think your students deserve it."
With that Sunderland concluded his presentation.

Don McGehee, an evolutionist and the science supervisor for the State Department of Education, spoke briefly
against the bill.

"Creationism is generally a religious belief," he said. He
added that introducing it into the public schools would
violate the principle of the separation of church and state.

He noted that the National Association of Biology
Teachers, the National Association for the Advancement of
Science and the Council of State Science Supervisors had
written position papers opposing the bill.

It amazed me that McGehee could sit through Sunderland's masterful presentation and not hear a word he said.

That day, after hearing McGehee's testimony, I discovered that evolutionists' minds are completely closed on the
subject of origins. That also amazed me for I thought scientists were supposed to consider all the information available
to them and with an open mind. I suspicioned we would
have a lot of opposition in the days ahead from the State
Department of Education.

By the time the debate ended most of the Education
Committee members had left the room. Since there was no
longer a quorum, and the committee could not vote on the
bill, no action was taken and the chairman dismissed the
meeting.

We were disappointed but resolved that day to continue
the fight.

A week later I went back to the committee and asked
them to pass a study resolution on the subject of balanced
treatment concerning origins. It passed unanimously. A
subcommittee was appointed to hear testimony during the
next nine months when the legislature was not in session
and to report their findings back to the legislature.

Our next hearing was held in November and things really started heating up.

During my opening remarks, I attempted to establish the nature of any forthcoming bill regarding balanced treatment of creation-science and evolution-science. The act would:

• Protect academic freedom by providing student choice.

• Guarantee freedom of belief and speech.

• Prevent the establishment of religion.

"In the Supreme Court decision in 'Torcaso vs. Watkins' in 1963, the court declared that secular humanism is a bonafide religion," I told the committee. "The number one tenet of secular humanism is the theory of evolution."

We tried to establish early in the debate that if the evidences for creation-science were considered to have a religious basis — which we did not believe — then the decision regarding secular humanism in "Torcaso vs. Watkins," should prove that there would be the same basis for saying evolution also has a religious basis. However, we contended that if only the scientific data of the two theories were presented in the classroom, completely apart from any religious references, that it did not constitute the establishment of religion.

I also told the committee it was grossly unfair to teach only one concept of origins and censor out the other.

"Clarence Darrow in his closing argument before the Scopes trial in an earlier generation said it is absolute bigotry to teach only one view of origins," I told the subcommittee members.

I knew the members of the panel had been exposed to the numerous newspaper articles which inaccurately called creation-science "religion" and evolution "science."

I also read a letter which I had received from Judge Braswell Dean, chief judge of the Court of Appeals in Georgia, concerning the constitutionality of creation-science. The letter said:

"Having been an adjunct professor of constitutional law, having lectured on origins, creation and evolution from a legal viewpoint, and having studied and having researched this legal issue for many years, it is my belief that the passage of your bill to teach scientific creationism or scientific evolution are not only constitutional, but failure to teach either one without the other is, in my opinion, placing the government and the school board in an unneutral position, which would be unconstitutional. Teach both or teach neither, and you have neutrality and constitutionality."

During my remarks, I also informed the committee that two large school districts — Dallas, Texas, and Tampa, Florida — recently had instituted the balanced treatment in their schools. And, I explained that they had encountered no serious problems with the implementation of both theories of origins.

Dr. Edward Boudreaux, an internationally known chemist from the University of New Orleans, was the first to speak in behalf of creation-science. He has served at the university for 18 years and has expertise in organic and theoretical chemistry, astrophysics, cosmology, biochemistry and biophysics.

"What I want to present to you today is science," Dr. Boudreaux told the subcommittee members. "As a scientist I am interested in the truth and my interest in creation-science is in the interest of science."

Dr. Boudreaux then explained the origins of evolution as presented by Charles Darwin in *The Origin of Species.*

"It was not very scientific," he said. "But it had a purpose — to initiate and foster a humanistic philosophy."

In order to understand why science would become involved in fostering a humanistic philosophy, people must understand something about the social, religious and political climate in Charles Darwin's day, Dr. Boudreaux explained.

"There was economic unrest and social unrest," he said.

"So they were ready for something new."

He said that science was making great strides during that era and the door was open for Darwin's new theory.

"That was the band wagon to get on," he said.

Dr. Boudreaux then proceeded to a very scholarly discussion of the subject of entropy.

"We have a law in nature which can be observed and validated through physical and chemical evidence," he said. "It says that when left to itself matter will tend spontaneously in a direction of disorder.

"When a person dies, decay gradually sets in and it goes on to more complex biological components. The end result is one of greater disarray.

"What is the issue here? The issue is that if you are going to produce something according to the evolutionary hypothesis, it has to be produced from a less perfect form. You must go contrary to the law of entropy."

His argument solidified the creationists' argument that when physical changes take place the tendency is downhill, not uphill, as the evolutionists teach.

Dr. Boudreaux then charged that evolutionary teachers were brainwashing children in the public schools.

"We have a duty to give the truth to all levels — from grade school through college — and present it objectively," Dr. Boudreaux said. "That is our duty as scientists and as educators."

Dr. Charles Harlow, professor of engineering at Louisiana State University in Baton Rouge, was the next speaker in favor of giving equal time to creation-science.

He said parents who have children in public schools should study the subject of evolution to be able to counteract the teaching. He added that most parents aren't aware of how the teaching affects their children.

"All our experience tells us that one species will not change into another," Dr. Harlow said. "All our natural experience tells us that an ape does not change into a man. No one has ever observed anything like that. . . . "

Dr. Harlow said that evolution is little more than a conjecture about what happened.

"I personally would prefer that neither idea be taught because we do not know what happened . . . and we will never know," he said. "So all we are doing is putting out conjectures that we cannot substantiate. But if they do insist on teaching evolution, then as a parent, the alternative (creation) should be presented."

Dr. Harlow said the theory of evolution has great philosophical and psychological influences on school children and even affects the way they live their lives.

"The exhuberance to promote evolution has created a number of frauds," he said. "At the turn of the century the name of the game was to find your own 'missing link.' "

The next speaker on behalf of creation-science was Nick Kalivoda, director of communications for Louisiana State University in Baton Rouge. He said his purpose for testifying before the panel was based on his interest in the freedom of information.

"This meeting is being held, I think, to determine whether children will benefit from freedom of information or if they will be handicapped educationally by having little or no information about creation," Kalivoda said. "There can really be no freedom for man if he has no choices. Demagoguery and those political systems which we find most offensive to the dignity of man are threatened by information which might expose their weaknesses."

Kalivoda pointed out that if evolution were really scientific, its proponents would welcome competition so that the truth would be more evident by comparison with other views.

He pointed out that evolutionists have a monopoly on science teaching as it relates to origins. He also said that if a graduate student in a state university told his professor he believes in creation-science it would be very difficult to get an advanced degree.

"The professors have a built-in protection," he said. "They multiply their own kind."

Kalivoda also said it was unfair for evolution scientists to allow only one view to be presented.

"Our refusal to hear the other side implies that all scientists agree, so our children come out of schools believing all scientists believe in the evolution theory," he said. "That is certainly not the case and we do them a disservice to allow that kind of thing to go on."

Kalivoda said it is reprehensible for anyone who claims to be a scientist or educator to limit information that attacks his view.

"Our school system seems to want to limit the information because they are threatened by the creation view," he said.

Tom Moore, a high school principal and biology teacher, also made a very enlightening presentation to the subcommittee.

"I am acutely aware of what it is like to try and teach a biology class using a textbook that uses the origin of living things and the origin of man from a one-sided viewpoint," he said. "As a science teacher, I have to do more than just dissiminate facts to students. My primary goal is to teach my students to think scientifically . . . I must teach them to think as a scientist thinks . . . there are two important things that I have to get across to them.

"First of all, they have to learn how to observe and, second, they have to learn how to experiment.

"That is the basis for all scientific knowledge. We use our five senses to observe and we draw conclusions on what we observe. We manipulate variables and experiment to determine relationships. Those two things are very important. . . . Those are the basics for scientific knowledge. If you go beyond observations and if you go beyond experimentations, then you are beyond the realm of science.

"Scientific knowledge can only go so far as the things you can observe and things you can experiment with. This is why it is difficult for me to teach the theory of evolution because origins — such as the origin of man, the origin of matter, the origin of energy, the origin of the earth, the origin of the universe — any origin is beyond the realm of science. Any explanation of how man got here or how the animals got here or how the earth got here cannot be scientific. That enters into another field of study — historical study.

"What happened a million years ago is a matter of history, not a matter of science."

Moore also discussed the law of biogenesis which says that only life can produce life, a law which has been used to dispute the evolutionary hypothesis.

"We know from observation and experimentation that life produces life," he said. "That is the law of heredity. But this law of heredity is in no way related to the theory of evolution.

"The principle of biogenesis is that the life of every organism comes from its parents.

"Does life ever spring from non-living matter? We can find no evidence of this happening. So far as we can tell, life comes from life.

"Biologists call this the principle of biogenesis. . . . In other words, through observation and experimentation, we have determined that life comes from life."

The biology teacher then explained what evolutionists say about how life began on earth. He said they believe that billions of years ago the earth was quite different from today. The atmosphere was made up of gases such as ammonia, methane and water vapor. There were elements such as carbon, hydrogen, nitrogen, oxygen and others.

"Energy from lightning will split some molecules," he said, and that's how the evolutionists say life began.

"That is spontaneous generation," he said. "Living matter arriving from non-living matter."

Moore said the concept of spontaneous generation contradicts the law of biogenesis.

"They cannot be put together," he said.

"From what we know, based on scientific facts, and what we know through observation and experimentation, the theory of evolution is an impossibility," he said. "Other forces that we don't know about must have been at work there if evolution occurred. Situations that we know nothing about must have been present for evolution to occur.

"So what we have is not really a theory. . . . I object to the idea that it is taught as a theory. There are certain limitations to the formulation of a scientific theory and evolution does not meet the criteria. Evolution is based on certain postulates that do not fit observation and experimentation, that we know about, and can never be proved. These postulates must be accepted if evolution were to take place. We must accept those assumptions blindly, such as the idea of spontaneous generation."

Kay Reiboldt, a Shreveport housewife and mother, represented the Pro Family Forum and spoke in favor of creation-science in the classroom.

"We support the idea of openness as a principle of science," she said.

"In recent years, the cosmos, the solar system, and life on earth have been presented in the classroom only from an evolutionary point of view. . . . The creationist point of view has been overlooked.

"We are concerned with the social, ethical, and scientific consequences of the current exclusive presentation of the humanistic evolutionary point of view regarding origins. The students have a right to know there is an alternate creationist point of view. They have a right to know the scientific evidences which support that alternative. They have a right to make a choice between evolutionary and creationist points of view relating to understanding of origins and science."

Mrs. Reiboldt also said parents have some rights which relate to what is presented to their children in the classroom.

"The primary goal of the parent and the upbringing of children was reaffirmed by the U.S. Supreme Court in 1972," she said. "It stated the history and culture of western civilization brought a strong condition of parental concern for the nurture and upbringing of their children.

"The primary role of the parent and the upbringing of their children is now established as an enduring American tradition.

"Also, in December of 1974, the U.S. Commissioner of Education, Terrell H. Bell, stated, 'Parents have the right to expect the school and their teaching approaches and selection of instructional material to support the values and standards children are taught at home. And if the school cannot support these values, they must at least avoid deliberate destruction of them.'

"It would be wrong to say that the biological theory of evolution has gained universal acceptance among biologists or even among geneticists. This is perhaps unlikely to be achieved by any theory which is so extraordinarily rich in philosophies and humanistic implications. But its acceptance is so broad that the proponents (of creation-science) complain of the inability to get a hearing of their views.

"It is particularly important that there be openness in the classroom relating to matters of origins," she said.

"Consideration must be given to the fact that there is a growing acceptance of the creation-science theory," she said. She added that many respected scientists in America view creation-science as a valid approach to origins.

She said the Pro Family Forum strongly advocates "a policy of openness as a principle of science with respect to presentations of matters of origins in the classroom.

"We are not asking that religion be presented in the classroom but we are asking that scientific evidences supporting both of these alternative points of view be presented," she said. "We call upon all of the school boards, manufacturers of textbooks, concerned citizens, and teachers of educational agencies to resist pressures towards scientific dogmatism and presentation of only one point of view of origins."

During the various other subcommittee hearings an abundance of scientific data was presented to support creation-science.

We also mentioned that many of the great scientists of history were creationists.

Among them was Dr. Wernher von Braun. Following is a statement by von Braun which he made in 1972, before his death:

> For me, the idea of creation is not conceivable without invoking the necessity of design. One cannot be exposed to the law and order of the universe without concluding that there must be design and purpose behind it all. In the world around us, we can behold the obvious manifestations of an ordered, structured plan or design. We can see the will of the species to live and propagate. And we are humbled by the powerful forces at work on a galactic scale. And the purposeful orderliness of nature that endows a tiny and ungainly seed with the ability to develop into a beautiful flower. The better we understand the intricacies of the universe and all it harbors, the more reason we have found to marvel at the inherent design upon which it is based.
>
> While the admission of a design for the universe ultimately raises the question of a Designer (a subject outside of science), the scientific method does not allow us to exclude data which lead to the conclusion that the universe, life and man are based on design. To be forced to believe only one conclusion — that everything in the universe happened by chance — would violate the very objectivity of science itself. Certainly there are those who argue that the universe evolved out of a random process, but what random process could produce the brain of a man or the system of the human eye?

Some people say that science has been unable to prove the existence of a Designer. They admit that many of the miracles in the world around us are hard to understand, and they do not deny that the universe, as modern science sees it, is indeed a far more wondrous thing than the creation medieval man could perceive. But they still maintain that since science has provided us with so many answers the day will soon arrive when we will be able to understand even the creation of the fundamental laws of nature without a Divine intent. They challenge science to prove the existence of God. But must we really light a candle to see the sun?

Many men who are intelligent and of good faith say they cannot visualize a Designer. Well, can a physicist visualize an electron? The electron is materially inconceivable and yet it is so perfectly known through its effects that we use it to illuminate our cities, guide our airlines through the night skies and take the most accurate measurements. What strange rationale makes some physicists accept the inconceivable electrons as real while refusing to accept the reality of a Designer on the ground that they cannot conceive Him? I am afraid that, although they really do not understand the electron either, they are ready to accept it because they managed to produce a rather clumsy mechanical model of it borrowed from rather limited experience in other fields, but they would not know how to begin building a model of God.

I have discussed the aspect of a Designer at some length because it might be that the primary resistance to acknowledging the "Case for Design" as a viable scientific alternative to the current "Case for Chance" lies in the inconceivability, in some scientists' minds, of a Designer. The inconceivability of some ultimate issue which will always lie outside scientific resolution should not be allowed to rule out any theory that explains the interrelationship of observed data and is useful for prediction.

We in NASA were often asked what the real reason was for the amazing string of successes we had with our Apollo flights to the moon. I think the only honest answer we could give was that we tried to never overlook anything. It is in that same sense of scientific honesty that I endorse the presentation of alternative theories for the origin of the universe, life and man in the science classroom. It would be an error to overlook the possibility that the universe was planned rather than happened by chance.

Some of the members of the subcommittee appeared shocked to learn that evolution was so deeply entrenched in the public school curricula. They assured us that serious consideration would be given to the information presented to them in the hearings.

CHAPTER 3

Evolutionists Fight Back

One morning the evolutionists in Louisiana woke up and discovered they had a problem on their hands. Their neat little monopolistic educational playhouse was about to come tumbling down.

They had controlled the evolutionary teaching of origins in the state for half a century and they determined to fight any attempts to break that monopoly.

They began their counterattack strategy maneuvers by using the ploy of name calling.

They called me personally an "Ayatollah," "charlatan," and "fascist." They referred to creation-science as "trash," "bogus," "sham," and "pseudo-science."

Their sticks and stones sometimes nearly broke our bones but their words really didn't hurt us. Their ignorance did.

During the days following the final passage of the Balanced Treatment Act, they mounted an intensive effort to stop Governor Dave Treen from signing the bill into law.

The capitol news corps engaged in a lot of wishful thinking by predicting the governor would veto the bill.

Editorial writers tried to intimidate him by saying he would send Louisiana back into the dark ages if he signed it into law. Others suggested he would lose a lot of political support if the bill became law.

The Shreveport Journal on July 16, 1981, wrote:

"Creationism belongs at the top of the bad bill list because it threatens academic freedom and blurs the separation of church and state. . . . The hope is that Treen will veto the bill and spare the state further embarrassment over this throwback to the infamous monkey laws of the 1920s."

But the governor was undaunted by all the threats. And when he signed the bill into law you could hear the evolutionists scream all over the state.

During a news conference the governor gave his reasons for signing the bill into law. He said:

"It has been argued to me that academic freedom will be harmed by the inclusion of instruction in creation theory. I find this argument curious," the governor said.

"If the purpose of education is to persuade young minds to a specific point of view, then competing thought should be excluded from educational curricula. If, on the other hand, the purpose of education is to expose young minds to a variety of ideas and concepts, then, of course, differing points of view should be presented.

"Academic freedom can scarcely be harmed by inclusion; it can be harmed by exclusion."

The editorial writer of *The Times* in Shreveport commented on the governor's actions by writing:

"It's fairly apparent that even Treen, who expressed doubts about his decision to sign the bill, found himself faced with a bill on which clear majorities of both the House and Senate had given approval — the supposed will of the people — and he, as governor, had no solid reasons to veto it. That's the thing about the creationists: They may not convince you to be for it, but they leave you little room to be against it. That's the barrel they've built, and they have succeeded in putting a lot of people over it. . . .

"What is now to be seen is whether or not the semantics will play in court. The American Civil Liberties Union is already challenging 'scientific creationism' in Arkansas, the only other state with a law providing for its teaching, and has said it will challenge Louisiana's law. The grounds will be constitutional — that church and state do not mix.

"What the ACLU will have to prove, though, is that the term 'scientific creationism' is a semantic sham, and that the body of facts it represents is, in fact, religion by any other name. That may not be as easy as you'd think. You may believe that it's just a masquerade for teaching religion in the public schools — and frankly, I do — but proving it, given the terminology set forth by creationism and its advocates, is another matter entirely.

"It is that terminology, most likely, which led Governor Treen to remark last week that when the ACLU lawyers read the Louisiana bill, they may want to drop their court challenge here. The language is carefully chosen and constructed to avoid mention of religion, or the imposition of religious beliefs in the classroom. Again, tricky. And politically, very shrewd. The people who put this movement together are not amateurs; anyone who thinks they are is woefully underestimating them."

Next, the evolutionists mounted a letter-writing campaign — letters which were gladly printed by evolutionist newspapers.

Critics of the law made the following statements in letters to the editor, news lines and public appearances:

"They (creationists) are attempting an end run around the separation of church and state by simply calling 'creationism' scientific. Giving it the name scientific doesn't make it science. Creationism is a religious doctrine and shouldn't be taught in the public schools," Dr. Douglas A. Roston, science professor, Louisiana State University in Baton Rouge.

"Religion and science are mutually exclusive and serve different purposes to society. The society which cannot make the separation shall have neither good science nor good religion," Dr. William Craig, professor of science, University of New Orleans.

Dr. George King, representing the Louisiana Academy of Sciences, said creation is a religious teaching which asks people to say, "I believe." But, he said, evolution is a science which leads people to say, "I conclude."

Dr. Douglas Hously of the University of New Orleans said creationist writings "misquote, quote selectively and quote wholly out of context."

The *Baton Rouge State-Times* editorial writer apparently was so upset about the law that he nearly came apart at the seams. He wrote the following editorial:

> According to a legend of the Mohawk Indians, the creation of man was a process of trial-and-error. When Sat-Kon-se-ri-io the Good Spirit, had finished making the animals, birds and other creatures, he finally decided to make a creature that would resemble himself.
>
> He fashioned a little clay man, and set it in the fire to bake. But he fell asleep, and when he awoke, he rushed to the fire and removed the little clay man. But his little clay man was burnt black. According to the Mohawks, this little man was the first Negro.
>
> The Good Spirit tried again, with a fresh piece of clay. But the river sang a sleepy song, and again he fell asleep. He awoke and removed the little man. But he had slept only a little while, and the man was half-baked. This, say the Mohawks, was the first white man.
>
> Again he tried, with a lump of perfect clay, and this time stood by the fire. When it was done, he removed it. It was just right — a man the red color of the sunset sky. It was the first Mohawk Indian.
>
> This is an Indian version of creationism, and it was based on observations the Mohawk believed to be true and accurate. It might be considered, then, the Mohawk version of scientific creationism.
>
> The Mohawk legend seems an appropriate example to illustrate one of the basic flaws of legislation approved by the Louisiana Legislature this week. . . .
>
> The legislation is, to return to the Mohawk example, half-baked, in the sense that it orders a study only of evidence that supports the Judeo-Christian belief — the version of creation according to the book of Genesis — without considering whatever evidence there may be to

support the beliefs of American Indians, Greeks, Romans, the Norse, Hindus and Buddhists of how the universe and life on earth began.

Joe Cordill, Jr., of Shreveport wrote the following letter to the editor of *The Times:*

I consider myself a pretty smart guy. I am a high school graduate. I subscribe to *Reader's Digest, People* magazine and the *National Enquirer.* All of the guys on my bowling team think I'm a pretty witty guy. You can ask them.

I'm not really religious. I think that's mainly because I never could pronounce Deuteronomy and I couldn't spell Leviticus. I still take my family to church on Sundays and I have a kid who sings in the choir. She sang a solo last week. . . ."

Anyway, I've been reading in the paper where Louisiana Senator Bill Keith is trying to make the schools teach the kids that this here evolution is not the only theory of how the world started. I don't really understand what "creationist" is. I know it has something to do with God. I don't really understand evolution either. I know it has something to do with monkeys.

Last week, I won an "eat dinner with a famous psychic" contest sponsored by a magazine. Our family had a famous psychic over to our house. The psychic sent us a sealed envelope before he came which we opened right before we ate. Would you believe it? We had meatloaf to eat, just as he had predicted. He also bent some of our silverware by just looking at it. (Although my wife, Betty, said that our disposal had done that a few weeks ago.)

But the best thing the psychic did was tell us how he believed the earth was created. He told us that the world was created as a high school science fair exhibit on the planet Rutabagus. I forget the date but it was sometime before 1837. I forget the creator's name, but I think it was Bubba.

Bubba created the earth in five days. He had to because the project was due on Friday and because of his religion, he couldn't work on weekends. His planet didn't have weekends anyway — only a five-day week.

Bubba gave earth a seven-day week so when he

visited he would have two extra days. He figured he could save money on air fares and car rentals on the weekend.

Bubba thought he had won first prize in the science fair because he had done such a good job. Then the judges found out that his parents had helped him. They had redesigned various parts of South America, Africa and Istanbul. Bubba was disqualified for cheating. He decided to sabotage the project. He put salt in the oceans, hid most of the oil under the deserts and made it real cold in Cleveland in the winter.

I think this theory explains just about everything. I can understand it anyway, and my kids seemed to understand it. Would you please pass this on to Senator Keith so that others can have this taught to them. I'll get him the psychic's address but don't expect him to answer smarty pants questions like how was Bubba created.

Richard L. Uznanski, of Lafayette, Louisiana, wrote the following letter to the editor and it also appeared in *The Times* of Shreveport:

"As a scientist, I applaud Senator Bill Keith, the Louisiana Legislature, and Governor Dave Treen for enacting a bill which allows the Biblical myth of creation to be discussed in public schools. Creationism dominated western thought on the origins of the earth and of life for thousands of years. Increasing scientific knowledge showed creationism to be false by 1800, but the myth was kept alive by religious fundamentalists. Creationism could not be discussed in public schools because of the doctrine of separation of church and state. Louisiana students finally have a chance to learn what is wrong with the Biblical story. . . . "

H.H. Frenkle wrote the following to the *New Orleans Times-Picayune:*

" 'Monkey Trial' Treen has torn his pants. Your placing the story of the Scopes trial next to the news item was clever.

"If Governor Treen goes along with the belief that we were created by God, I can readily go along with the statement that a thousand years are in His eyes like a single day.

I can readily believe that even a million years do not mean any more to Him than does a single day to a mortal. Thus there is no difficulty in accepting both Genesis and evolution as different sides of the coin.

"But why would the governor say the bill does not mandate the teaching of religion? The bill plainly states 'shall' give balanced treatment. Does the governor think 'shall' means 'why?' This is playing with the truth, and there is a word for that.

"If the governor does not readily believe in a literal interpretation of Genesis, the word for his signing is hypocrisy.

"Either way, Treen has not covered himself with glory."

The most surprising letter written by the evolutionist critics of creation-science was written by Frank Stagg, senior professor of New Testament Interpretation, Southern Baptist Theological Seminary, Louisville, Kentucky. His letter appeared in the *Times-Picayune* and said:

> The actions of the Louisiana Legislature and governor concerning "scientific creationism" appall me. There are so many things wrong with such action that it is difficult to conclude where the most damage is done. That religion and government thus encroach upon one another seems apparent. The implication of conflict between Genesis and unimpeded science overcome only by dogmatic science and dogmatic religion is a disaster. For government to become the lackey of reactionary religion is a disservice to both religion and government.
>
> The theory that God made a young earth (not more than 10,000 years old!) to look old (instant antiquity!) invites the deplorable idea that God cannot be trusted. Probably the most serious damage is to those perceptive enough to see in such action the compromise of authentic religion, government and science. So-called "scientific creationism" invites skepticism, not faith.
>
> "Scientific creationism" does not represent a credible understanding of the intention of Genesis any more than a credible science. Genesis does offer a credible understanding of existence as a divine origin and as answerable to God, with the human being seen as a

creature made in the image of God, a part of creation yet more, like God and yet not God. Genesis sees the human being as moral, created in awesome freedom, with the power to opt for life in creative relationship with God, with other people and with the rest of creation or to go it alone at the price of self-destruction.

Genesis does not intend to be a textbook in science. Science may find its commission in Genesis, where mankind is commissioned to occupy, enjoy and exercise dominion over the rest of creation. This leaves room for honest investigation, a function native to a healthy human being.

Every time religion employs political or ecclesiastical power to impede science or impose its dogmas upon people it loses. Just now the church is embarrassed as it tries awkwardly to rectify its 1633 injustice to Galileo, whom it forced to compromise his own integrity by recanting his finding that the earth moves around the sun and not vice versa. Greater even than its sin of obscurantism was religion's shameful act in forcing a sick and aged man to falsify himself and his scientific finds. To ask public school teachers and their students to carry on the fiction that "scientific creationism" is science and not religion (and thus not an encroachment of religion and state upon one another) at least borders upon the crime against Galileo.

Robert F. Scott, in a letter to *The Times* of Shreveport, wrote:

"Can any taxpayer justify in this day and time the expenditure of millions of dollars to legislate religious dogma in our schools? Surely not. . . . Kindly . . . leave me and my tax dollars out of it."

Bert Watkins also wrote a letter to *The Times:*

"The creationist books I have read are completely useless as sources of scientific ideas. Don't drag our school children back into the 19th century."

Thomas Jukes, a University of California biochemist, said during a symposium on creation-science at the Superdome in New Orleans:

"Creationists have a right to their own beliefs, but what

they're trying to do is terrorize school teachers.

"There is a growing alarm in the scientific community that creationism's attack on science will spoil scientific education for the young."

Joan Allen wrote the following to the *Times-Picayune* in New Orleans.

> The teaching of scientific creationism in our public schools would be an unconscionable act of exploitation by our religious ayatollahs. Everyone knows that once a child is taught the "God" fear, religionists then have a "patsy" for life, someone they can control and exploit. It is very hard for a child to cast off this fear and do his own thinking once it has been implanted early enough. Religionists know this; that is why they are trying to get their foot in the door of our public schools.
>
> There are far too many religious fairy tales being taught in the churches and homes of America. Our children need a safe haven where they can feel free to do their own thinking. A recent study by the University of Houston clearly shows that the children of Catholic and Baptist mothers are not as intelligent as the children of non-religious mothers.
>
> America, protect your children from the religious enslavers! Our public schools are the only haven they have left. If that should tumble, our children will have no safe place to be free from the religious enslavement of their minds.

Two professors collaborated on a letter to the *Times-Picayune* in New Orleans in which they paraphrased the famous words by the great German patriot Martin Neimoller.

William Cohen, Ph.D., professor of biochemistry, and Manie K. Stanfield, Ph.D., assistant professor of biochemistry, both of Tulane Medical School, wrote:

"First the self-righteous, self-declared Moral Majority censored TV shows; I didn't speak up because I rarely watch TV.

"Then they tried to abolish legal abortions and circumvent the Supreme Court ruling; and I didn't speak up because it didn't affect men.

"When they insisted on prayers in the schools despite the constitutional requirement for separation of church and state, I didn't speak up because it didn't seem important enough.

"Now that they are forcing the teaching of religion in my science classes, there is no one, not clergy, not politicians, nobody to speak up for me."

Figaro, a weekly magazine in New Orleans, captioned a four-page article on creation-science: "Evolving Ignorance."

"Science is in trouble in Louisiana," *Figaro* said. "There is a movement to give creationism equal credibility with evolution in school textbooks. . . .

"This scientific sleight-of-hand introduces what is essentially a religion concept — entirely unverifiable — into the objective world of science."

These are only a few examples of the thousands of letters the evolutionists wrote as they mounted an offensive against creation in the classroom. Since only about 15 percent of the people in Louisiana oppose the balanced treatment, there were so many opposing letters it made me wonder if every evolutionist in the state may have written at least two letters to the editors of some newspaper.

Creationists answered them and tried to explain that creation-science is pure science and as unreligious as evolution. Science and not Genesis. Science and not the Old Testament. But most evolutionists' minds are completely closed because their belief is based on faith and not on facts.

The most formidable and outspoken critic of creation-science was State Superintendent of Education Kelly Nix.

Nix learned that many of the school teachers organizations — composed primarily of teachers trained in humanistic universities — opposed the law.

Opposition to creation-science started with the biology

teachers, spread to other science teachers, then to the teachers organizations and ultimately to the state superintendent. Since most teachers disliked the superintendent, he saw a great political opportunity to make some points with them by opposing the law.

Nix fired off a volley at creation-science and the legislature during a speech in Bossier City, Louisiana. He said it disappointed him when the legislature passed the Balanced Treatment Act. He then argued that elected officials "should never mandate courses."

Moreover, he said, the legislature voted no money to implement the law and astounded everyone present by saying it would cost $8 million to establish creation-science in the classrooms of the state.

That speech to Bossier Parish educators clearly outlined the strategy the evolutionists would use to break down support for the new law. Let's examine that strategy:

1. Should the legislature ever mandate courses in the public schools?

I assigned the senate attorneys to research for me how many times the legislature had passed laws mandating curricula.

Their research revealed that on numerous occasions, during the past 10 years, the legislature had given instructions to the schools on certain courses they must teach. Courses on patriotism and free enterprise were mandated. The legislature attempted to mandate sex education but it was voted down by the representatives and senators. Legislative acts also set up standards regarding testing, minimum standards and a host of other requirements for the schools.

However, no one, including Nix, said much about those mandated courses. But when the creation-science law passed they argued that the legislature was meddling in school affairs.

Who owns the schools? The public, of course. Who represents the public? Legislators.

Modern-day educators want the public to stay out of the school's business and perhaps that's what's wrong with public education today.

They want us to take our little children to the school door, leave them there and from then on they belong to the school. Parents are urged to attend PTA but not to tell educators how or what to teach their children.

2. Will it cost millions to implement the Balanced Treatment Act?

Let me ask: How much did it cost to implement evolution?

It was interesting that Superintendent Nix harangued creation-science before the Bossier Parish teachers. Two years earlier their school board had voted to implement equal time for creation-science in all the science classrooms in the parish. Obviously Nix did not know that when he delivered his anti-creation, pro-evolution speech to them.

Bossier Parish received no outside state funds to implement the balanced treatment. They already had discretionary funds, for the purchase of library books and textbooks, which were available to them to help defray the costs for balanced treatment.

Those teachers must have been amazed to hear Nix say it would cost $8 million statewide. It actually cost them very little.

One study in Arkansas showed creation-science would be implemented there for about $30,000. That would involve using mimeographed and xeroxed materials and color slide presentations.

But Nix — and other evolutionists — hammered away at how much they thought it would cost the taxpayers. That was the first time I ever knew of him being concerned about keeping costs down.

Their distortions of the truth failed to move the people of Louisiana away from their staunch position that creation-science should receive equal time.

3. Will the little school children suffer loss in

other areas of academia if so much money is spent on implementing creation-science?

The liberal *Shreveport Journal* parroted Nix's charges that the costs for implementation would be exorbitant.

"State Education Superintendent Kelly Nix and his science supervisor, Don McGehee, believe implementation will divert between $5 million and $7 million from the state's basic education programs — reading, writing and arithmetic . . . ," the *Journal* wrote in July of 1981. "Just to place one so-called scientific creationism book in the library of each of the state's 1,448 schools will cost a minimum of $18,824," McGehee said. . . .

"The cost of providing the state's 28,857 upper elementary and secondary science teachers with just one copy of the textbook will run approximately $375,141. In-service training for those same teachers will cost another $857,610; teachers attending the training sessions will cost $264,000."

Let me comment that when Nix instituted the teaching of values clarification in the schools of the state, he didn't complain about the cost of in-service training for the teachers. Yet, because he opposes creation-science he kept talking about the exorbitant costs.

Let's take a look at the above article in the *Journal:*

1. The writer said it was the "so-called" creation-science theory. Was the writer writing a news story or an editorial?

2. Nix said reading, writing and arithmetic will suffer if creation-science is implemented. Louisiana students are suffering already with low test scores in these areas. Actually they are among the lowest in the nation.

The fear tactics employed by Nix are only a camouflage to hide the real issue of his opposition to the theory.

Take textbooks, for instance. If they buy certain textbooks on the subject of origins there will be no additional costs. Some textbooks present the balanced treatment. And when a few more states pass balanced treatment acts, some of the major textbook publishers will begin changing their approach to origins. Some already have begun to make the transition.

Before Governor Treen signed the bill into law, Nix said: "I hope he· vetoes it. I told the school superintendents this morning to get influential people back home to call and telegram him asking him to veto it."

A newspaper reporter called me, told me about Nix's statement and asked for a response. I issued the following brief statement:

"Kelly Nix, in all the years I've known him, has never known what he's talking about. This is just another good example of how far behind the times our superintendent of education remains. . . . He has worked behind my back to try to kill the bill. I'm glad he has finally come from behind his cloak of hypocrisy and shown us where he really stands."

Even though the media, the teachers organizations and the superintendent of education all strongly opposed and ridiculed creation-science, the people of the state were not swayed. They continue to stand for academic freedom, and against censorship, in the classroom.

CHAPTER 4

Hopeful Monsters

Since the days of Darwin, evolutionists have searched for the "missing link" that would resolve the question of evolution forever. Yet the phantom has eluded scientists and does so today.

Therefore, some leading evolutionary spokesmen, undaunted by their failure to produce even one "missing link," have changed their direction and emphasis. They have resurrected the old "hopeful monster" idea once proposed by Richard Goldschmidt.

Goldschmidt, a well-known geneticist, believed that mutations ordinarily produce monsters. But he speculated that occasionally some genetic changes might produce "hopeful monsters" whereby an ape could take a giant leap forward and suddenly become a man.

Stephen Jay Gould has written an article entitled: "The Return of the Hopeful Monsters." Gould, professor of geology and paleontology at Harvard, is one of the major spokesmen for evolutionary thought in America. He also is a self-proclaimed Marxist.

In the article, Gould wrote: "As a Darwinist, I wish to defend Goldschmidt's postulate that . . . major structural transitions can occur rapidly without a smooth series of intermediate stages."

Why would Gould make such a statement? Because he and other evolutionists have been unable to find the "missing links" or transitional forms. So, they take the easy way out. They change the name of the game.

Gould now couches his belief in evolution in such terms as "punctuated equilibria," a fancy name for "hopeful monsters."

Debates between creationists and evolutionists historically have centered on the lack of "missing links" in the fossil record.

Why would the evolutionists make such a radical change in their thinking? For decades they have asserted that Darwinian evolution is credible. Suddenly they no longer believe it themselves.

So, enter "hopeful monsters."

A historic meeting took place in the Field Museum of Natural History in Chicago in 1980. Some 160 of the world's leading paleontologists, anatomists, geneticists and biologists attended the meeting. The majority of those present acknowledged they believed in some form of "punctuated equilibria" or "hopeful monsters."

The shift from traditional evolution to "punctuated equilibria" apparently was the result of frustration. However, rather than accept what creationists have told them for more than 100 years — that they never would locate the missing phantoms — the evolutionists just shifted gears and pressed on toward their goal of proving the unprovable.

Newsweek Magazine on November 3, 1980, wrote: "Evidence from fossils now points overwhelmingly away from the classical Darwinism which most Americans learned in high school: that new species evolve out of existing ones by the gradual accumulation of small changes, each of which helps the organism survive and compete in the environment. Increasingly, scientists now believe that species change little for millions of years and then evolve quickly, in a kind of quantum leap. . . ."

The new evolutionists are quick to point out that they are not sure whether the quantum leap occurs in a hundred years or many thousands, even millions of years. They say their change in thinking does not agree with the creation scientists who for 100 years have been telling them their theory was wrong. They insist that "punctuated equilibria" is a valid explanation of origins.

"The new theory, according to paleontologist Steven Stanley of Johns Hopkins, draws a crucial distinction between two kinds of evolution: gradual, small changes within a species ('microevolution'), and sudden, gross changes that mark the emergence of a new species ('macroevolution')," *Newsweek* reported. "The former is a specialized case of Darwin's theory of natural selection. . . . But where Darwin, from observations begun in the Galapagos Islands, concluded that enough small changes would eventually create a new species, the revised theory holds that a new species arises by some different mechanism — perhaps even in a gross random mutation in a single generation."

Newsweek continued its report:

> This is the theory of "hopeful monsters," a point of bitter contention among geneticists and biologists. To some geneticists all monsters are hopeless. Such a major change in structure can only be the result of gross chromosome rearrangements. So many other delicate systems would be set awry as a result that the organism could not survive.
>
> But the significance of hopeful monsters, if they exist, is that they seem to flout the law of natural selection. They are subject to it in a general sense: better monsters will over the long run drive out weaker ones. But the mutation need not represent an advance in fitness: a mutant gene can spread throughout the population even if it carries no particular survival value as long as it is not markedly harmful. The iron line of Darwinism — that each new species represents an advance in fitness over its predecessor — seems to have been breached. . . .
>
> It is no wonder that scientists part reluctantly with Darwin. His theory of natural selection was beautiful in its simplicity and has served well for over a century. To tamper with it is to raise a host of questions for which there are no answers. The new theory also raises the

troubling question of whether man himself is less a product of 3 billion years of competition than a quantum leap in the dark, just another hopeful monster whose star was more benevolent than most.

It fascinates me that such cataclysmic changes are taking place within evolutionary thought. But it amazes me that an incorrect form of evolution — compared to current thought — is being taught in school classrooms in Louisiana and throughout the land.

If we are going to teach evolution, we should teach current thought and not some fairy tale that even the evolutionist no longer accept. Yet that is exactly what is happening in our schools today.

Tens of millions of school children have been indoctrinated in Darwinian evolution. Will anyone ever try to correct the error? Will anyone go to them and say, "We were wrong"?

For more than 100 years the evolutionists have preached Darwin's doctrines. Now it has crumbled. It is incredible to think that the evolutionists would be so unscientific.

What do they do about it? They cop out.

How could such a thing happen? Why would the evolutionists continue living in their dream world? What will be their next approach to evolution when creation scientists disprove "punctuated equilibria" and "hopeful monsters"?

Their erratic thought behavior is based on their religion or faith and they will never give up in their attempt to prove evolution is true.

Robert Jastrow, Ph.D., well-known professor of astronomy and geology at Columbia University in New York, provides us with some clues as to why the evolutionists continue their hopeless quest to prove an unprovable theory.

"Perhaps the appearance of life on the earth is a miracle," he wrote in *Until the Sun Dies,* one of his many scientific books which have been published. "Scientists are reluctant to accept that view, but their choices are limited;

either life was created on the earth by the will of a being outside the grasp of scientific understanding, or it evolved on our planet spontaneously, through chemical reactions occurring in non-living matter lying on the surface of the planet.

"The first theory places the question of the origin of life beyond the reach of scientific inquiry. It is a statement of faith in a Supreme Being not subject to the laws of science.

"The second theory is also an act of faith. The act of faith consists in assuming that the scientific view of the origin of life is correct, without having concrete evidence to support that belief."

Professor Jastrow, an evolutionist and agnostic, is a forthright man. Apparently he is big enough not to be threatened by the fact that evolution is based on faith and cannot be proved.

But our school children never hear such statements. They are told that evolution is science rather than faith.

The transition from Darwinism to "hopeful monsters" is based on faith. That faith rules out creation so must rely on an alternative view. And they will do anything necessary to try to prove that view.

Let's take a look at Charles Darwin to better understand how we arrived at where we are today.

He published *The Origin of Species* in 1859 in an attempt to prove that man evolved from lower forms of animals and had a common ancestry with apes.

On December 27, 1831, Darwin boarded the H.M.S. Beagle for a five-year, 40,000-mile trip around the world. He was a naturalist and would represent the Geological Society without pay.

After some time, the Beagle reached South America and sailed to the Galapagos Islands near Ecuador. There, Darwin observed several giant tortoises and noticed that some of them had shells which differed from those on the other islands. He carefully recorded the differences in his journal along with notes about plants, animals and reptiles he had observed.

Darwin filled many notebooks with his observations and when he returned to England five years later, he used the notes for a massive work on the subject of natural selection.

During the next 30 years he spent most of his time poring over his dozens of notebooks and writing. He was convinced that human beings and all other life slowly evolved from lower forms.

His ideas were quickly grasped by many of the people of his day, particularly those who had intellectually discarded the concept of creation as unscientific.

Through the years most of his followers lost interest in his theory because there was no proof for it. During that time Darwin himself said: "I am in a hopeless muddle concerning the origin of things. Our ignorance of the derivation of things is very profound. I must be content to remain agnostic."

However, as he lay dying, he embraced the concept of creation. He also reflected on his life work, saying: "I was a young man with unformed ideas. I threw out queries, suggestions, wondering all the time over everything; to my astonishment the ideas took like wildfire. People made a religion out of them."

Darwin died 100 years ago this year.

Although Darwin recanted his view of natural selection and evolution from lower forms, thousands of scientists have appropriated his works in their search for the understanding of origins.

The faith aspect of evolution should be indelibly written in the minds of every parent in this land. This faith — based on the false teaching of evolution — is being forced on school children, camouflaged as science, but, in fact, religion.

The Pro Family Forum has published a pamphlet entitled: "Are Evolutionists Like Three Blind Mice?"

How are they like three blind mice? "They can't see the past! They can't see the present! They can't see the evidences!" the pamphlet said.

"If evolution really happened in the past, it should be a continuing process, observable today — or at least during recorded history," the pamphlet said. "In other words, there should be some proof of evolution going on in life around us."

The Pro Family article then asked:

• "Have you seen any apes moving into houses and getting jobs?"

• "Have you seen any fish sprouting legs lately?"

• "Have you seen a snake growing wings to become a bird?"

"Sound silly? Of course! You know that all fish are still fish, apes are still apes, snakes are still snakes, and man is still man."

The article continued: "Even the frog, the evolutionists' best example of development from fish to amphibian, is no example at all. The tadpole may look like a fish, but it grows into a frog every time. Further, the fish they have chosen to evolve into an amphibian, supposedly evolved into a lizard, not a frog! . . . changes have been made within species (such as many different dogs — large and small, etc.), but the most brilliant scientists have never been able to cross the species (such as to cross a dog with a cat or a horse with a kangaroo). All effort to do so produces an animal which cannot reproduce.

"As long as man has been recording history (approximately 6,000 years), no evidence of evolution has ever been found. SURELY, SOMEWHERE, SOMETIME, SOME CHANGES MUST HAVE BEEN OBVIOUS, IF EVOLUTION IS TRUE!

"Theistic evolutionists believe that the geologists have proved the earth to be billions of years old, so the Biblical six days of creation must be wrong. What they don't realize is that geologists have based their method of dating rocks on the fossils found in them. So, the rocks are dated according to the theory of evolutionary development! In other words:

"THE ROCKS ARE DATED BY EVOLUTION, THEN USED TO PROVE EVOLUTION!

"Evolutionists claim that the proof of evolution lies in the fossils, and geologists claim that the age of the fossils proves the age of the rocks. . . .

"Even though NO evolutionists were present at the beginning (and neither were the rest of us), they take it on themselves to tell us that life happened by accident.

"Imagine, if you will, that someone told you that a jet airplane just came into existence by accident. You'd tell him to have his head examined!

"Yet, evolutionists tell us to believe that all forms of life — unbelievably complex — just happened by accident!

"We know that every living thing is made up of cells. Your body contains over a million million of them. It would take about 40,000 blood cells to fill this letter O. Yet each cell contains tiny structures, each with a job to do. Cells 'breathe,' take in food, get rid of waste, and reproduce. Each cell contains a master plan, called the nucleus, that controls everything it does.

"The master plan is called DNA, and consists of a chemical code. DNA makes every living thing different from every other living thing: makes a dog different from a fish, a rose different from an ape, and is one of the things which makes an ape different from a man.

"If evolution is true, then someone, or something, had to break the DNA code, and drastically change it — not once, but THOUSANDS OF TIMES. The fact is that scientists have not broken the DNA code today, and no evidence is found of any changes in the code at any time in the past.

"Could these complex cells, which are the simplest forms of life, be just 'freak accidents' of nature?

"Could these carefully designed marvels have simply evolved over millions or billions of years?

"COMMON SENSE SAYS 'NO!'
"SCIENTIFIC EVIDENCE SAYS 'NO!'

"The points mentioned above are only a fraction of the abundant evidence which confirms, even for children, that evolution is both unreasonable and unscientific.

"Yet, for several decades this evidence has not been given to students in our public schools," the article concluded.

Let's take a look at some of the supposed intermediate forms or "missing links" which for many years have been used to support Darwinian evolution.

Heidleberg Man. When certain remains of a so-called Pleistocene creature were found near Heidleberg, Germany, it was hailed as a great discovery and victory for evolution. Artists drew pictures of what they thought he must have looked like. They actually made him appear to be half man and half ape. Those pictures appeared in national magazines, school textbooks and museums.

But close examination by reputable scholars discovered that the Heidleberg Man's jawbone was that of a human.

So, the Heidleberg Man was a hoax.

Nebraska Man. Paleontologists digging in Western Nebraska found a tooth. None of them had ever seen a tooth quite like it. So, they speculated it must be the tooth from a "missing link." Again, artists drew pictures of what the Nebraska Man must have looked like — based only on the contour and size of one tooth.

Pictures of the Nebraska Man appeared in magazines, textbooks and were displayed at the Smithsonian Institute in Washington, D.C.

The evolutionists were so sure they had found a "missing link" that testimony about the Nebraska Man was entered as evidence in the Scopes Trial.

Some years later a chemist asked to examine the tooth. He discovered it was the tooth of an extinct pig.

As one of my friends said, "The evolutionists tried to

make a monkey out of the pig, but the pig made a monkey out of the evolutionists."

Piltdown Man. When scientists discovered some bones in Piltdown, East Sussex, England, they believed the bones to be from a very primitive man. They based their speculation on the shape of the skull fragments.

Webster's Dictionary says the Piltdown Man was "uncovered in a gravel pit at Piltdown and used in combination with comparatively recent skeletal remains of various animals in the development of an elaborate fraud."

Some evolutionists will go to great lengths to try to find proof for their theory.

Peking Man. The remains of this creature were found in a cave in Choukoutien, China. Yet, somehow, all the remains have disappeared. The evolutionists undoubtedly didn't want another hoax on their hands.

Neanderthal Man. Pictures of this man show him walking almost upright, but with distinct ape-like features.

School children by the tens of millions have been told he was a true "missing link."

However, during the Congress of Zoology in 1958, Dr. A.J.F. Cave said the skeleton of the man, which had been discovered in France 50 years earlier, was that of an old man who had a bad case of arthritis. But true man — not ape, or half man and half ape.

Do you realize that vast segments of the public still believe Neanderthal Man to be a true "missing link"?

New Guinea Man. He was also considered an unquestionable intermediate form and a proof for evolution. But anthropologists have recently discovered a species of people — exactly like the New Guinea Man — still living north of Australia. And guess what? They, too, are true men.

Cro-Magnon Man. He was discovered in southern France. Again, the evolutionists said he was a true "missing link." The creature was tall and erect, just like men. He later was determined to be a man.

There have been others but none have stood the test of

scientific examination.

Therefore, the evolutionists, realizing these so-called "missing links" were making monkeys out of them — came up with the "hopeful monster" theory. But this concept is even more incredulous than traditional Darwinism.

Let's look at some statements by great scientists who recognized evolution to be only a theory and not fact.

CHARLES DARWIN: "Long before the reader has arrived at this part of my work, a crowd of difficulties will have occurred to him. Some of them are so serious that to this day I can hardly reflect on them without being in some degree staggered; but, to the best of my judgment, the greater number are only apparent, and those that are real are not, I think fatal to the theory."

Toward the end of his life, Darwin openly admitted: "Not one change of species into another is on record. . . . We cannot prove that a single species has changed into another." Darwin, Charles, *My Life and Letters,* Vol. 1, page 210.

THOMAS HUXLEY said that "evolution was not an established theory but a tentative hypothesis, an extremely valuable and even probable hypothesis, but an hypothesis none the less." Himmelfarb, Gertrude, *Darwin and the Darwinian Revolution,* Doubleday and Co., New York, 1959, page 366.

PROFESSOR THEODOSIUS DOBZHANSKY, a leading spokesman for evolution, has said that "it would be wrong to say that the biological theory of evolution has gained universal acceptance among biologists or even among geneticists." Dobzhansky, Theodosius, *Science,* Nov. 29, 1963, page 366.

DR. AUSTIN H. CLARK, noted biologist of the Smithsonian Institute, stated: "There is no evidence which would show man developing step by step from lower forms of life. There is nothing to show that man was in any way connected with monkeys. . . . He appeared SUDDENLY and in substantially the same form as he is today. . . . There are no such things as missing links."

He also said, "So far as concerns the major groups of animals, the creationists appear to have the best of the argument. There is NOT THE SLIGHTEST EVIDENCE THAT ANY ONE OF THE MAJOR GROUPS AROSE FROM ANY OTHER. Each is a special animal complex, related more or less closely to all the rest, and appearing therefore as a special and distinct creation."

Meldau, Fred John, *Witness Against Evolution,* Christian Victory Publishing Co., Denver, Colo., 1953, page 39, 40, 73.

PROFESSOR G.A. KERKUT, an evolutionist, states, ". . . There is the theory that all living forms in the world have arisen from a single source which itself came from an inorganic form. This theory can be called the 'GENERAL THEORY OF EVOLUTION' and the evidence that supports it is not sufficiently strong to allow us to consider it as anything more than a working hypothesis." Kerkut, G.A., *Implications of Evolution,* Pergamon Press, Elmsford, N.Y., 1961.

JOHN T. BONNER, who reviewed Kerkut's book for *Science* (Vol. 133, March 17, 1961, page 753), "This is a book with a disturbing message; it points to some unseemly cracks in the foundations. One is disturbed because what is said gives us the uneasy feeling that we knew it for a long time deep down but were never willing to admit this even to ourselves. It is another one of those cold and uncompromising situations where the naked truth and human nature travel in different directions. The particular truth is simply that we have no reliable evidence as to the evolutionary sequence of invertebrate phyla. We do not know what group arose from what other group or whether, for instance, the transition from Protozoa occurred once, or twice, or many times. . . .

"We have been telling our students for years not to accept any statement on its face value but to examine the evidence, and, therefore, it is rather a shock to discover that we have failed to follow our own sound advice."

PROFESSOR ALBERT FLEISHMAN, professor of Comparative Anatomy at Erlangen University, said, "The theory of evolution suffers from grave defects, which are becoming more and more apparent as time advances. It can no longer square with practical scientific knowledge, nor does it suffice for our theoretical grasp of the facts. The Darwinian theory of descent has not a single fact to confirm it in the realm of nature. It is not the result of scientific research, but purely the product of imagination." Fleishman, Albert, *Victoria Institute,* Vol. 65, pages 194, 195.

SIR WILLIAM DAWSON, Canada's great geologist, said of evolution: "It is one of the strangest phenomena of humanity; it is utterly destitute of proof." Dawson, Sir William, *Story of Earth and Man,* page 317.

DR. ROBERT A. MILLIKAN, famous physicist and Nobel prize winner, said, "Everyone who reflects believes in God." Millikan, Robert A., *The Commentator,* June 1937.

In an address to the American Chemical Society, he said: "The pathetic thing about it is that many scientists are trying to prove the doctrine of evolution, which no scientists can do."

LOREN EISLEY, a leading evolutionist, says: "With the failure of these many efforts, science was left in the somewhat embarrassing position of having to postulate theories of living origins which it could not demonstrate. After having chided the theologian for his reliance on myth and miracle, science found itself in the unenviable position of having to create a mythology of its own: namely, the assumption that what, after long effort could not be proved to take place today, had in truth taken place in the primeval past." Eisley, Loren, *The Immense Journey,* Random House, New York, 1957, page 199.

DR. D.M.S. WATSON writes, ". . . The theory of evolution itself is a theory universally accepted not because it can be proved by logically coherent evidence to be true but because the only alternative is special creation, which is clearly incredible." Watson, D.M.S., "Adaption," *Nature,* Vol. 124, 1929, page 233.

DR. W.R. THOMPSON, a world renowned entomologist, was for many years the director of the COMMONWEALTH IN-STITUTE OF BIOLOGICAL CONTROL at Ottawa, Canada, and was selected to write the foreword to the new edition of Darwin's *Origin of the Species.* In that foreword he made the following very significant statement: "As we know, there is a great divergence of opinion among biologists, not only about the causes of evolution but even about the actual process. This divergence exists because the evidence is unsatisfactory and does not permit any certain conclusion. It is therefore right and proper to draw the attention of the non-scientific public to the disagreements about evolution. But some recent remarks of evolutionists show that they think this is unreasonable. This situation, where men rally to the defense of a doctrine they are unable to define scientifically, much less demonstrate with scientific rigor, attempting to maintain its credit with the public by the suppression of criticism and the elimination of difficulties, is abnormal and undesirable in science." Thompson, W.R., *Introduction to Origin or Species,* by Darwin, E.P. Dutton & Co., New York, 1956.

DR. GEORGE WALD, a Nobel prize winner, chooses to believe in evolution even though he said he regards it as a scientific impossibility. He says, "The only alternative to a spontaneous generation is a belief in supernatural creation. . . . " Wald, George, "Innovation and Biology," *Scientific American,* Vol. 199, Sept. 1958, page 100.

SIR JAMES JEANS, eminent British astronomer, wrote, "We discover that the universe shows evidence of a designing or controlling power . . . from the intrinsic evidence of His creation, the Great Architect of the Universe now begins to appear as a pure Mathematician." Jeans, Sir James, *The Mysterious Universe,* MacMillan, New York, 1947.

SIR AMBROSE FLEMING, president of the British Association for the Advancement of Science, stated, "Evolution is baseless and quite incredible." Fleming, Sir Ambrose, *The Unleashing of Evolutionary Thoughts,* 1935, page 254.

DR. WERNHER VON BRAUN, who masterminded the V-2 rocket of Germany in World War II and the space program of the United States for two decades, said in a speech at Taylor University: "The idea of an orderly universe is inconceivable without God — the grandeur of the cosmos confirms the certainty of creation. One can't be exposed to the law and order of the universe without becoming aware of a divine intent."

DR. WARREN WEAVER, formerly chairman of the board of the American Association for the Advancement of Science, said, "Every new discovery of science is a further 'revelation' of the order which God has built into His universe." Weaver, Warren, *Look Magazine,* April 5, 1955, page 30.

FRED JOHN MELDAU wrote, "In nature we find endless variety within the species of genus; but absolutely NO CHANGE from one family to another. Summed up in four words, the laws governing all life prove there are MUTATIONS BUT NO TRANSMUTATION, which simply means that there are many varieties within any group, but there can never be one 'kind' of life mutating (changing) into another family." *Witness Against Evolution.*

PROFESSOR T.H. MORGAN says, "Within the period of human history we do not know of a single instance of the transformation of one species into another." Morgan, T.H., *Evolution and Adaptation,* page 43.

RICHARD GOLDSCHMIDT, Ph.D., professor of zoology, University of California, said, "Geographic variation as a model of species formation will not stand under thorough scientific investigation. Darwin's theory of natural selection has never had any proof . . . yet it has been universally accepted. There may be wide diversification within the species . . . but the gap (between species) cannot be bridged. . . . Sub-species do not merge into the species either actually or ideally."

These great men of science speak for themselves. Why is their message ignored by contemporary evolutionists?

"Hopeful monsters" is weaker science than Darwinism, but unless things change, our school children will be hearing about them both for many years to come.

CHAPTER 5

The Religion of Humanism

Evolution is pure humanism.

A very wise man once described secular humanism as "human ego and intelligence gone mad."

Those who embrace this religion try to make a man out of the Creator and a creator out of man. They say there is no such thing as a Creator, that the idea was devised by man. In other words, the Creator didn't create man — man created the Creator. Their faith is in man and no one else.

Experts who study the religion of secular humanism estimate that there are fewer than 300,000 hard-core humanists in this country. But millions believe in some of the tenets of humanism and tens of millions are being influenced by their religion every day.

Though limited in numbers, their influence is far-reaching and permeates every facet of society. For instance, they:

• Control public education in America today.

• Wield a strong influence on the news media.

• Influence most all textbooks used in our public schools.

• Dominate many areas of state and federal government, particularly the programs governmental bodies carry out.

Their humanistic doctrines are flooding this country today. It has a terrifying impact on the people in general and school children in particular.

The U.S. Supreme Court (Torcaso vs. Watkins, 1961, and United States vs. Seeger, 1964) declared that humanism meets all the criteria for a bonafide religion.

Mel and Norma Gabler of Longview, Texas, have studied secular humanism for many years. The following is based on their findings. Most of the teachings come directly from *Humanist Manifestos I* and *II,* the humanist's bible.

1. **Evolutionary dogma.** Humanists falsely state that it is a scientific fact, rather than a theory.

2. **Self autonomy.** This is the belief that there is no higher authority and therefore everyone, including children, must become their own authorities.

3. **Situation ethics.** This concept teaches that nothing is absolutely right or wrong and that everything is situational or relative.

4. **Christianity is passe.** It negates Christianity and all reference to the supernatural.

5. **Sexual freedom.** This idea provides the impetus for public sex education — but with no moral teaching included. It makes fun of modesty, purity, chastity and abstinence. It encourages abortion, premarital sex and homosexuality.

6. **Total reading freedom.** They believe children should have the right to read anything and everything and that parents should not be a part of the decision-making process.

7. **Death education.** This concept teaches there is no life beyond the grave.

8. **Internationalism.** World citizenship, according to humanists, is far superior to national citizenship, patriotism and love for country.

9. **Socialism.** They believe that state ownership of property is far superior to private ownership.

This man-made religion is particularly widespread in public school textbooks.

Textbooks dealing with evolution teach that man is

just an animal with no soul.

Let's look at several examples:

". . . Infants can grasp an object such as a finger so strongly that they can be lifted into the air. We suspect this reflex is left over from an earlier age in human evolution, when babies had to cling to their ape-like mothers' coats while mothers were climbing or searching for food." *Understanding Psychology,* Random House.

"Man is, without question, the most outstanding product of evolution. In a sense, human evolution has been in process since the first stirrings of life on earth." *BSCS Molecules to Man,* Houghton-Miffin.

This is pure speculation based on evolutionary, humanistic faith. Since they deny the existence of a Creator, they try to fit scientific data into their own presupposition that all things evolved — molecule to man.

Now let's look at textbook examples of the belief in self autonomy. This is the idea that man is just an animal and responsible to no one.

"The place, the opportunity, and their bodies all said 'Go!' How far this couple goes must be their own decision." *Masculinity and Femininity,* Houghton-Miffin.

This teaching is directly opposed to the teaching most children learn in the home.

"From whom might you resent getting some unasked-for advice about how to dress, how to wear make up, or how to behave? Why? (From some teachers, from 'old-fashioned' parents, from bossy older brothers and sisters.) . . . " *Rebels and Regulars,* Teachers Manual, MacMillan.

Do you see the irony in this kind of teaching?

Parents, through their taxes, construct school buildings, purchase textbooks and pay teachers' salaries. Yet often the textual material attempts to undercut parental authority.

What about situation ethics?

"There are exceptions to almost all moral laws, depending on the situation. What is wrong in one instance may be right in another. Most children learn that it's wrong to lie. But later they may learn that it's tactless, if not actually wrong, not to lie under certain circumstances." *Inquiries in Sociology,* Allyn & Bacon.

Here's another example.

"Note (to teacher): Please refrain from moralizing of any kind. Students may indeed 'tune out' if they are subjected to 'preachy' talk about 'proper English' and the moral obligation to 'do one's best' in class and to 'lend a hand' to the underdog in a battle. . . . " *Gateway English,* MacMillan.

How does secular humanism negate traditional teachings of Judeo-Christian belief on which this nation was founded?

"Anthropologists studying human customs, religious practices, ritualism, and the priestcraft came to the conclusion that men created their own religious beliefs so that the beliefs answered their special needs. The God of the Judeo-Christian tradition was a god worshipped by a desert folk . . . and heaven was high above the desert. . . . " *Perspectives in U.S. History,* Field Educational Publications.

This is a frontal attack on the beliefs of many school children, beliefs they learned in the home.

Why is such a blatant attack on the beliefs of the children necessary? Because the humanists must destroy all other religions for their own religion to thrive and capture the minds of all the people.

The Humanist Manifesto says: "Traditional theism, especially faith in the prayer-hearing God, assumed to love and care for persons, to hear and understand their prayers, and to be able to do something about them, is an unproved and outmoded faith. Salvationism, based on mere affirmation, still appears as harmful, diverting people with false hopes of heaven hereafter. Reasonable minds look to other means for survival."

Here's another example of how humanism attempts to destroy other systems of belief.

"A great many myths deal with the idea of rebirth. Jesus, Dionysus, Odin, and many other traditional figures are represented as having died, after which they were reborn, or arose from the dead. . . . " *Psychology for You,* Oxford.

These textbooks, written by humanistic writers, offer no alternative views other than humanism. Other views are not tolerated.

What do humanists teach about sexual freedom?

Here's what they believe and teach your school children.

"There are some who adopt more permissive standards for themselves and others. They propose conditions outside of marriage under which they feel that sexual relations should be permitted." *Psychology for Living,* Webster/McGraw.

Opinion polls reveal that most Americans still believe sex outside of marriage is wrong. So why is it being presented to school children as right and acceptable? To spread humanistic thinking.

What about this?

"Contrary to past belief, masturbation is completely harmless, and in fact, can be quite useful in training oneself to respond sexually. . . . " *Life and Health,* Random House.

Harcourt's *Study of Human Relationships* recommends that the students study the "advantages" of group marriage.

Save the Schools Newsletter recently reported that "there is a continuing movement underway in American education to de-sensitize prospective teachers about the subject of homosexuality and to make any form of sexual license and perversion a routine matter for classroom discussion."

Humanists recommend total reading freedom for children. Perhaps that's why school libraries have so many books on witchcraft, black magic and the occult.

Here's what *The Humanist Magazine* said about reading freedom:

"Something wonderful, free, unheralded, and of significance to all humanists is happening in the secondary schools. It is the adolescent literature movement. They burn *Slaughterhouse Five* in North Dakota and ban a number of books in Kanawah County, but thank God the crazies don't do all that much reading. If they did, they'd find that they have already been defeated. Adolescent literature has opened up Pandora's box. . . . Nothing that is part of contemporary life is taboo in this genre, and any valid piece of writing that helps to make the world more knowable to young people serves an important humanistic function."

Humanists have outsmarted parents. While parents slept, humanists were busy writing school textbooks which struck at the very heart of the beliefs most parents had instilled in their children.

Now let's examine the subject of death education, an integral part of the humanists' approach to social engineering of school children into the humanistic image. This teaching includes visits to morgues, the writing of personal epitaphs and the study of suicide. It teaches there is no life beyond the grave and therefore there should be no moral restraints here on earth.

Here are some textbook examples of death education.

"Dying As An Orgasmic Event — The thought of death sometimes occurs in a sexual context . . . in the event of orgasm, like the event of dying, involves a surrender to the involuntary and the unknown." *Life and Health,* Random House.

Also, the humanists promote the teaching of euthanasia or mercy killing in the textbooks.

"Many a theoretically strong pro-life stance melts into a belief in euthanasia as soon as one is confronted with a loved one who is screaming in agony or lying in a comatose state. . . . " *Life and Health,* Random House.

This textbook discussion of euthanasia closely parallels the *Humanist Manifesto II.* Note the similarities:

"There is no credible evidence that life survives the death of the body. We continue to exist in our progeny and in the way that our lives have influenced others in our culture. The individual must experience a full range of *civil liberties* in all societies. It also includes a recognition of an individual's right to die with dignity, euthanasia, and the right to suicide."

This all means that our school children are learning values in the classroom through the study of humanistic textbooks that are diametrically opposed to the values of the home. And the taxpaying parents are paying for this indoctrination.

Now let's examine the humanistic idea of internationalism. This is the idea that loyalty to the world should replace national patriotism and love for one's own country.

"A third alternative future would occur if the global political system moved in some entirely new direction, in other words if the entire global system changed. What if national and multinational corporations faded out of existence, to be replaced by one single conglomerate? . . . " *American Government: Comparing Political Experiences,* Science Research Associates.

Have you ever wondered why so many school children never pledge allegiance to the flag of the United States? Humanists discourage it for it promotes patriotism.

Rather, in some textbooks, the United States is condemned, criticized and ridiculed while communist countries and their leaders are glorified. And the taxpayers pay the bill for it all.

Public school textbooks also promote socialism or the idea that state ownership of property should replace the free market system.

Humanist Manifesto I makes these socialistic goals quite clear.

"The humanists are thoroughly convinced that existing acquisitive and profit-motivated society has shown itself to be inadequate and that a radical change in methods, controls and motives must be instituted. A socialized and cooperative economic order must be established to the end that the equitable distribution of the means of life be possible. The goal is a free and universal society in which people voluntarily and intelligently cooperate for the common good. Humanists demand a shared life in a shared world."

Textbooks parrot this humanistic thinking about socialism.

"By 1936 the depression was seven years old, and conditions were steadily worsening. Only one bright spot stood out in the gloom: The collectivist economy of the Soviet Union . . . stood without unemployment, universal hunger, or signs of disease. . . . Meantime, Americans live with the system they have, patching it here and there. . . . " *Perspectives in U.S. History,* Field Educational Publications.

One textbook also eulogizes Mao Tse Tung, the leader of the Chinese revolution, as one of the great men of our time. Yet they failed to mention that he and the Soviets butchered millions of innocent people, took away their freedom and dissolved property rights.

That's humanism in action!

How did the humanistic religion spread so rapidly throughout our society?

Mel and Norma Gabler say it all started with the "humanistic/atheistic views" of John Dewey, the father of "Progressive Education" in this country. It spread through the "waves of educators who are proponents of Dewey's philosophies. . . . " Soon it engulfed teachers, colleges and the educational establishment in the land. Eventually it infected millions of students in the schools and overflowed to the general public.

According to the Gablers, the public is continually bombarded by humanistic philosophies, but most people in this country still have not rejected their traditional moral values.

Secular humanism has made its greatest inroads into society by influencing people in education, government and the news media.

Recently the Connecticutt Mutual Life Insurance Company released a report on "American Values in the '80s." It focused on the impact of belief as it relates to morality in this decade.

Their findings were astonishing. They interviewed a broad cross-section of the American public and here are some of their findings.

On the subject of abortion, the poll asked: "Is abortion morally wrong or is this not a moral issue?"

Some 65 percent of the general public said they believed it is morally wrong. But only 35 percent of those in the news media believed abortion to be wrong, only 29 percent in government and only 26 percent of the educators.

Another question in the poll asked: "Is homosexuality morally wrong or is it not a moral issue?"

Among the general public, 71 percent answered that homosexuality is morally wrong. However, only 38 percent of the news media, 36 percent of government officials and 30 percent of the educators think it is wrong.

Another question asked: "Are pornographic movies morally wrong or is it not a moral issue?"

Sixty-eight percent of the general public believes pornographic movies are wrong but only 50 percent of the educators, 47 percent of government officials and 46 percent of the news media.

Another 50 percent of the educators don't believe pornography is even a moral issue, the poll revealed. And, 53 percent of government officials and 54 percent of the news media agree with the educators that pornography is not a moral issue.

It is apparent that the secular humanists have been successful in winning control of most newspeople, government officials and educators. These people represent three of the most powerful groups on earth capable of molding the opinions of millions of people — particularly school children — every day in this land.

After reading Connecticutt Mutual's poll, it no longer surprised me that the news media, educators and government bureaucrats were the strongest opponents of the concept of equal time for creation-science in the classroom. Their orientation is secular and the idea of creation-science is a grave threat to them. Therefore, they vigorously oppose it.

They want only their opinion taught in the classroom. The thought of a Creator is a great threat to them. Educators, news media personnel and government bureaucrats working together have maintained a humanistic monopoly over the classroom for decades.

But the dike is just about to break and creation-science may prove to be the catalyst that returns public education back to the people where it belongs.

Humanists have been preaching their religion in the public schools and getting away with it. And there's no room for creation-science.

Do you know the argument the humanists use against teaching creation-science in the classroom? They say it is religion. That's sheer hypocrisy. Creation-science is just as unreligious as evolution-science.

Dr. Henry Morris, the eminent director of the Institute for Creation Research in San Diego, California, wrote an article in the *Impact Series* entitled "The Religion of Evolutionary Humanism and the Public Schools."

Here's what Dr. Morris says about it:

"The modern creationist movement and the resistance of secular educators to this movement have brought into clear focus one very important fact. Our American public schools and secular universities are controlled by the religion of evolutionary humanism. Furthermore, through its pervasive influence on the graduate schools and the textbook publishers this powerful concept has had significant impact even on Christian schools."

Another article written by Dr. Morris entitled "Evolution Is Religion, Not Science" makes it clear that evolution is religion disguised as science.

"Evolutionists often insist that evolution is a proved fact of science, providing the very framework of scientific interpretation, especially in the biological sciences," Dr. Morris said. "This, of course, is nothing but wishful thinking. Evolution is not even a scientific hypothesis, since there is no conceivable way in which it can be tested.

"For example, two leading evolutionary biologists have described modern neo-Darwinism as 'part of an evolutionary dogma accepted by most of us as part of our training. . . .'

"In view of the fundamentally religious nature of evolution, it is not surprising to find that most world religions are themselves based on evolution. It is certainly unfitting for educators to object to teaching scientific creationism in public schools on the grounds that it supports Biblical Christianity when the existing pervasive teaching of evolution is supporting a host of other religious philosophies. . . .

"In this perspective, it becomes obvious that most of the great world religions — Buddhism, Confucianism, Taoism, Hinduism, Animism, etc. — are based on evolution. . . .

"All of this points up the absurdity of banning creationist teaching from the schools on the basis that it is religious. The schools are already saturated with the teaching of religion in the guise of evolutionary 'science.' In the modern school, of course, this teaching mostly takes the form of secular humanism, which its own proponents claim to be a 'non-theistic religion.' It should also be recalled that such philosophies as communism, fascism, socialism, nazism, and anarchism have been claimed by their founders and promoters to be based on what they regarded as scientific evolutionism. . . . ''

The old argument that evolution is science and creation is religion is only a camouflage to keep the truth out of the classroom. That is censorship and an abridgement of academic freedom.

It is remarkable that evolution (a lie) is so strongly entrenched in the public school curricula. Yet creation-science (the truth) is censored by the humanists.

The only way to change such a travesty of justice is for the people to challenge secular humanism in the classrooms, in the halls of congress, in the state legislatures and before local school boards.

Here are some helpful tips on how to combat secular humanism:

1. Read your children's textbooks and become thoroughly knowledgeable on the contents of the books.

2. Determine the humanistic content.

3. Confront your child's teacher and instruct him/her that you don't want your child to be taught that which you consider secularistic or humanistic. Remember, the teacher works for you.

4. If you don't get satisfaction from the teacher, go to the principal.

5. If you don't get satisfaction from the principal, go to the school board.

6. Hopefully they will listen and try to do something about secular humanism in the schools. If they don't, wait until the next election and vote in a new school board that will be sympathetic with your concerns.

The primary tool of secular humanism in the public schools today is values clarification.

Concerning this tool for social engineering of school children, the Gablers wrote:

"For years, schools took a neutral stance on morals. Gradually, neutrality changed to attacks on moral standards, until student beliefs or values were under attack in both textbooks and school programs."

The method used to attack the traditional moral values of school children is values clarification.

Dr. Sidney Simon is the high priest of values clarification and here are some statements from his book *Values Clarification*, published by the Hart Publishing Company, Inc.

"Young people brought up by moralizing adults are not prepared to make their own responsible choices," Dr. Simon says.

He also urges teachers to help students get rid of the "immorality of morality."

And he suggests that teachers present values clarification in such a way that they will not get caught by the parents.

Here are some of the teachings and methods of values clarification which are being used to brainwash your children:

• Parents should not be allowed to make decisions for their children.

This is a direct attack on the role of parents in our society.

• Advice from parents is no more valuable than that which children receive from the New Left and the counter culture.

Again, by this teaching, parents are under attack.

• Patriotism is old-fashioned and outdated.

No wonder many schools have discontinued the pledge of allegiance to the flag.

• Social engineering of the children's minds toward

humanism is superior to the traditional values received through the home, church and constitutional ideas.

• Teachers should indoctrinate the children in values clarification by asking questions like: Do you approve of premarital sex for boys? For girls? Do you think sex education classes in the schools should include techniques for lovemaking, contraception? Would you favor a law that would limit the size of families to two children? Would you like to have different parents? Do you think we should legalize marijuana? Should we legalize abortion? Would you choose to die and go to heaven if it meant sitting around on a cloud and playing a harp all day? Should homosexuals be allowed to teach in the public schools? Do you approve of a couple trying out marriage for six months before getting married? Should we legalize mercy killings? What would you rather do on Sunday morning: sleep late, play with a friend or watch TV? Do you think parents should teach their children to masturbate?

We send children to school to learn reading, writing and arithmetic and some teachers find time to ask them if they believe the government should limit the size of families. No wonder the children can't read.

• Values clarification also makes fun of young men who go to war for their country; people who stress law and order; and those who do not believe in the free love or free sex concept.

That also is pure secular humanism and tens of millions of young people are receiving that indoctrination every day.

When we visualize the background, practice and goals of secular humanism in our society today we then can understand why those who embrace that religion are so violently opposed to giving equal time to the teaching of creation-science in our schools.

Secular humanism is the wisdom of man and it has no place for a Creator.

The Pro Family Forum issued a study entitled: "Is

Secular Humanism Molesting Your Child?" In the study, secular humanism is clearly unmasked and found to be a religion.

"Humanism is referred to by humanists as a 'faith' and a 'religion,' " the study reported. "Does this religion have effective Sunday Schools? Not exactly. It has effective Monday through Friday schools. . . . Our public schools . . . are rapidly changing from traditional education to 'change agents' for humanism. Who pays for it? You do."

One of the old, worn-thin arguments the secular humanists use on those who embrace traditional moral values is: "You're trying to impose your morals on us." This is the standard argument they use against creation-science.

But they are guilty of imposing their morality on us. Just look at what we've tolerated — the fruit of secular humanism:

Pornography . . . abortion . . . drunk drivers . . . proliferation of alcoholism . . . casino gambling . . . trash on television . . . rock and roll music that promotes Satan worship . . . occult books in school libraries . . . sex education without morality . . . homosexual teachers . . . values clarification . . . left-wing professors in tax-supported state universities . . . nihilism . . . situation ethics . . . music that glorifies sex, drugs and immorality . . . evolution.

We have been pushed far enough!

This detailed account of secular humanism will perhaps give you some insight into why the humanists are so vocal in their opposition to equal time for creation in the classrooms.

Creation-science drives a wooden stake into the heart of secular humanism. It can't long survive in the light of the truth.

CHAPTER 6

Tempest in the Typewriter

The news media is the public relations arm of humanistic thought in this land. And, to better understand why the news media has been almost universally opposed to equal time for creation-science in the classroom, it is important to understand why they take such a position.

Someone has said that we have three branches of government in this land — ABC, NBC and CBS. Although that may not be exactly true, every knowledgeable person knows the vast power the myriad news organizations have throughout America and around the world.

I was a newsman for 23 years and the city editor of the afternoon newspaper in Shreveport until 1979 when I resigned to enter the race for the Louisiana Senate.

During the last two years as a newspaperman I became quite disenchanted with my chosen profession. The reason for that disenchantment was that during the 1970s I saw the news media drift further and further to the left. Rather than applying an objective, unbiased approach to the coverage of the news, many news organizations began publishing or broadcasting the strong viewpoints of the owners, editors and even the reporters.

Some called it "The New Journalism."

When I graduated from journalism school in 1957, I had been well-schooled in the approach to my profession. The professors said to cover a story by trying to determine, as much as possible, exactly what happened, then write the story objectively and based only on the facts.

However, in the 1970s there was a trend toward more subjective reporting that included the ideas of the reporter or broadcaster on the subject. And many of the young broadcast and print journalists coming out of liberal universities had a lot of ideas implanted in their minds concerning social change in society. Little by little the reporter's own viewpoint became a strong factor in news coverage.

That opened Pandora's box.

For instance, if a reporter was pro-abortion, his stories ordinarily would reflect his bias. If the reporter were a homosexual, his articles would support that life-style and those who practice it. One homosexually inclined writer wrote a scathing story about some of the TV ministers who preach against homosexuality. Fairness would dictate that the writer declare himself to be a homosexual so the readers would clearly understand his bias. But such fairness does not exist in many newspapers.

During the last year I served as city editor of the afternoon paper in Shreveport, that once-proud newspaper wrote strong editorials favoring the new abortion clinic in the city; favoring the pornographic movie that had just come to town; and even editorialized in favor of people's right to drink alcoholic beverages in the city parks on Sunday.

On one occasion, a young writer attended a pornographic movie and wrote an explicit story on what he saw. The newspaper allowed it to be published in the weekend magazine section.

I had not seen the article and did not know it was in the magazine. When I arrived home that evening with the newspaper I handed it to my wife, Lowayne, and she began to read it. She was shocked when she read the article based on the young reporter's visit to the pornographic movie. That night she asked me never to bring the newspaper into our home again.

Many parents have begun to realize that so-called "family newspapers" often are filled with trash that should not be brought into the home. Some of the papers show lewd scenes in their advertisements of X-rated movies and even advertise for abortion clinics.

Recently the Shreveport police closed down the pornographic movies in the city. The police officers were acting within the limits of the law which unequivocally says the community has the right to set its own moral standards and the Shreveport community is overwhelmingly opposed to pornography.

The afternoon newspaper chastised the police for closing down the movies. The humanistic bias was clearly seen in the article which had the following headline: "Police Raids Give This Fellow the Blues."

"Unfortunately, I missed this month's weekly raid on the Glenwood and Capri theaters. But I wish I had been there. . . . It would have been reassuring to witness the salvation squad's attempts to save me from myself. . . . "

The writer then told how he went to a local video store and rented his own X-rated movie and took it home for a private showing. His point was: Why shut down the porno houses? A person can buy or rent the same films all over town.

The article poked fun at the police:

"While the legal system is busy trying to penetrate the obscenity quotient of the seized films, modern science and community standards allow me (for $3 plus deposit) to rent *Black Silk Stockings* — one of the confiscated films. And that's exactly what I did.

"Some politicians may not like the idea of me watching 'ultimate sex' acts . . . but then some politicians may engage in activities that I would find distasteful. . . .

— "Let's put aside the nonsense, the hypocrisy . . . I, for one, am d----- tired of the whole thing."

That particular newspaper had a conflict of interest in the matter. It carries advertisements for the porno movies.

Maybe that is one of the reasons it supports them so strongly.

Do you see what is happening? The owners and writers of the newspaper are trying to break down all moral restraint. Yet the U.S. Supreme Court has clearly stated that police can shut down porno movies if they feel the majority of the citizens disapprove of them.

It is very clear that any newspaper that allows filth in its news columns would be opposed to creation-science.

A South Louisiana newspaper decided to write a lengthy article on the creation-evolution controversy in the state. The reporter, a friendly sort of fellow, called and interviewed me for about an hour. I naively believed he would be fair.

The article turned out to be a personal attack on me — filled with innuendos, half-truths and outright lies.

Later I learned that the editors of the newspaper had held a meeting to decide how to deal with creation-science. They determined to make creation-science look like a joke and the author a fool.

There's nothing fair about that. But that's what we must expect from the news media which is eaten up with secular humanism.

The news media has a "bee-hive mentality." If one writer writes a negative article on a subject, all the others join him. This is dangerous for it cuts deep into the respected concept of objectivity and fairness in journalism.

The Balanced Treatment Act won an overwhelming victory in Louisiana. But stories on the proceedings gave three and four times as much news space to the few who opposed the bill as to the vast majority who supported it. Reading the newspaper accounts one could have believed the evolutionists won and the creationists lost.

We supposedly live in a pluralistic society. Yet the media seems to want to segregate creationists out of that pluralism. We are the new kids on the block and have really been pushed around by them.

People in the media no longer think like you do. Many of us have suspicioned this was true. Now it can be proven.

Public Opinion Magazine recently published the report of a survey of 240 members of the "media elite" including reporters, editors, columnists, bureau chiefs, news executives, TV correspondents, anchormen, producers, film editors and others.

The poll was conducted by Robert Lichter and Stanley Rothman who spent an hour each with the media leaders.

Paul Harvey said of this group: "These are the people whose efforts color what you hear and read of public affairs, particularly of President Reagan and his policies."

Here are some of the pollsters' findings:

• A mere eight percent are regular church-goers. Another 86 percent "seldom or never" attend church.

• Some 84 percent said adultery is okay.

• They believe they should run the nation.

One of the questions the pollsters asked them was: "Who directs American society?"

The media elite answered: business, the media and trade unions.

When asked: "Who do you think should run this country?" they ranked themselves first. Isn't that amazing? They certainly have high opinions of themselves yet have been elected to nothing by the people.

"These media people believe in redistribution of income, government-guaranteed jobs, affirmative action," Harvey said in a column.

He added that there is no way the Reagan administration can get its philosophy across to the American people "with such a preponderance of the media lying in wait to subvert and distort it."

We could say the same for creation-science. The media has done everything imaginable to distort the truth and subvert the facts.

But creationists need not fear the media. Apparently

they are speaking only to themselves and few others are listening.

Here's an example:

The Shreveport Journal conducted a poll on the subject of creation-science in June of 1980. The results surprised everyone, including me.

It revealed that 70 percent of the people in the Shreveport area favored equal time for creation-science, while only 17 percent opposed it and 13 percent had no opinion.

Apparently troubled by the results of the survey, the *Journal* launched an attack on creation-science. Editorials, opinion columns and slanted news stories ridiculed and made fun of the theory, trying to sway public opinion against it.

Six months later the newspaper, expecting to see a dramatic shift in public opinion on the issue, conducted another poll.

The second poll showed that 76 percent — or an increase of six percent — now favored creation-science. Some 15 percent opposed it and nine percent had no opinion.

More astounding was the fact that 56 percent of the people said they opposed teaching evolution in the classroom. Some 31 percent approved of teaching evolution and nine percent had no opinion.

They've run no more polls on the subject!

The news media has been very critical of the concept of creation-science but just remember that one good Creator is worth a thousand critics.

Creationists are becoming stronger every day all across America. Continued opposition can be expected from newspapers but their bias on the subject is just a tempest in a typewriter.

Most local television coverage of the issue has been quite fair. But there has been a distinct bias against creation-science manifested by the three major networks.

Let me show you the methodology a newspaper uses to slant public opinion toward a particular bias held by the publisher, editor or reporter — such as abortion or pornography — and to fight a concept such as creation-science. Here is the strategy:

1. Slant the stories written on the subject. There are many ways to do this. One of the easiest is just to give more space to the side with which the writer is in agreement. This method was employed routinely by newspapers in their coverage of creation-science. They almost always gave more column inches to the evolutionists than to creationists.

2. Ask questions which promote their own point of view. A skillful reporter can write stories mentally while asking questions. By the questions he asks, he can lead the person being interviewed down a primrose path and arrive at a conclusion which matches the reporter's own bias.

3. Use only selective quotes from the interview which fit the bias of the newspaper. This occurred numerous times in the creation-evolution debates. For instance, if a creation scientist presented some really strong scientific evidence for creation, the writers often would not print those arguments but something the man may have said that was totally extraneous to the question.

4. Leave out quotes with which the writer disagrees. This is one of the worst forms of bias.

5. Take quotations out of context to either diminish the statement's strength or pervert its meaning.

6. Place evolutionist stories in a good location on a good page of the newspaper where it can be easily seen by the readers, but bury the creation story back near the want ads. One of the Baton Rouge newspapers is very good at this.

7. Use more letters to the editors from evolutionists than from creationists. Also, select letters in such a way as to make evolutionists look like brilliant scientists and creationists look like religious fanatics.

One newspaper carried a letter to the editor from a

scientist opposing creation-science. The editors placed the letter in a good location on the page with a good headline.

Later, an internationally known scientist, who is a creationist, answered the above-mentioned scientist's letter giving scientific explanations for creation. But the newspaper refused to print the creation scientist's letter. Rather, the editors chose to print creationist letters from people who based their arguments on emotion rather than science.

8. Use pictures in such a way as to make evolutionists appear scholarly and creationists as buffoons.

Many of the cute little biased tricks used by newspapers also work well for the broadcast media yet are used to a much lesser degree.

The Associated Press (AP) and United Press International (UPI) have probably displayed the most rampant bias of all the news media.

Let me cite two examples.

The AP office in New Orleans writes or rewrites stories which are sent out to the newspapers of the state. Virtually all of the stories dealing with creation-science have said we believe the earth is only 6,000 years old.

I don't believe that and I'm the author of the creation-science law. Yet all of the AP and some UPI stories keep referring to the idea that the earth is only 6,000 years old and attach wording to that effect onto almost every news story they write.

On several occasions I tried to correct that misconception with the AP writers. I explained to them that creationists have differing views on the age of the earth. Some believe it is relatively young while others believe it is much older.

But the wire service continued to distort the creationist point of view by stating in each story that we believe the earth is only 6,000 years old.

Here's another example.

The Louisiana School Board Association met in Shreveport in February of 1982. During the meeting there was a lengthy and heated debate on creation-science. Both proponents and opponents spoke on the issue.

An AP story said the school board members voted to oppose creation-science. That was untrue since no action was taken on the matter. Yet the account of their opposition appeared in major dailies in the state and on numerous television broadcasts.

After carefully checking with school board officials concerning their action, it was determined that no action, pro or con, was taken.

I called AP in New Orleans and explained the facts to them and pointed out the error. The editor was insolent and said he didn't know that the story was wrong. When confronted with the facts, the editor said, "Well, I don't know if we'll do anything about it or not."

"I don't care what you do, but your story was wrong and I thought you might be interested in knowing about it," I replied.

As far as I know, AP never corrected the story and felt no obligation to do so.

Now let's look at how some specific newspapers dealt with the creation-science vs. evolution-science controversy in the state.

John McMillan, a columnist for the *Baton Rouge State Times,* wrote an article entitled: "Scientific Religionism." A careful look at his article reveals that apparently he never had read the bill, was biased against it and wanted to make fun of it. The article said:

> Learning that the Louisiana Legislature passed a bill requiring the teaching of "scientific creationism" in the public schools if those schools also teach the theory of evolution sent me scurrying to the calendar.
>
> After studying it awhile, I came to the conclusion it really is 1981.
>
> Creationism is a Bible-based theory of how the world was formed. It is called scientific for the same reason garbage collectors are referred to as sanitation engineers.

Following the terms of this legislation should present some interesting classes if the schools in Louisiana are going to continue teaching science. The science teacher would also have to teach creationism and that would probably make schizophrenics out of both the teachers and the students. As I see it, the teacher would first present scientific information and then the creationism theory.

Science: "The earth has been here billions of years, according to scientific examination, students."

Creationism: "The earth has been here for exactly 6,000 years, give or take a half hour."

Science: "The earth revolves around the sun, which is the center of our solar system."

Creationism: "And you want to know how the sun stood still in order to aid the Israelites? Well, in those days, students, the solar system worked somewhat differently."

Science: "There is no fish with a digestive system large enough to swallow a man whole, and if there was, a human being would die in short order under those conditions."

Creationism: "Men used to be smaller than they are today — little biddy suckers — and could hold their breath a long, long time."

Science: "The idea of a worldwide flood doesn't hold water, so to speak, because the water would have had to rise to a depth high enough to cover, say, Mount Everest, and if that had happened, the water would not have evaporated yet and we would not be here, and as far as I can tell, we are."

Creationism: "There was a worldwide flood. Where did the water go? Well, someone pulled the plug and it drained away. What happened to the people and animals? During the time of the flood, a farsighted man built a big boat and put two of every animal, reptile, bird and insect and every living thing aboard. How big was the boat? Have you ever been to Texas?"

Science: "Man is the only animal that has the power of speech."

Creationism: "Back during the time of the Garden of Eden, snakes could talk. They were very convincing when it came to advertising apples. Why can't they talk now? That's a subject for evolution class."

And now it's time for recess.

The Shreveport Journal, the strongest opponent of the bill of any news organization in the state, wrote an editorial severely criticizing Governor Dave Treen for signing the bill into law. The article, which follows, reveals the complete lack of understanding of the new law and the newspaper's own personal bias against it, a bias which led the editorial writers to color everything written about it. The article, entitled "Monkeying With Science," said:

> It took Governor Dave Treen a weekend of agonizing to reach the wrong decision on creationism. The reasons he presented Tuesday for signing the state's first monkey law were a poor argument for mixing science and religion by legislative fiat.
>
> Having admitted earlier that he himself does not interpret the Bible literally, Treen expressed reservations about the creationism bill's definitions and its ultimate effect. Yet like the legislature, he accepted at face value the sophistry that the bill only sought "balanced treatment" between two opposing "scientific" theories.
>
> Creationism is no more science than St. Peter was a geologist. Fundamentalist Christians have simply developed a new tack for getting around the U.S. Supreme Court, which ruled in 1968 that legislatures cannot ban the teaching of evolution.
>
> The modern monkey law, based on "creation science," makes no mention of God, nor does it say that the two Genesis accounts of creation are the preferred versions, out of many that exist in the world's religions. Be assured, however, that fundamentalists are not interested in having students learn that Hindus believe the universe rests on the trunk of an elephant whose legs are supported by four giant sea turtles.
>
> Senator Bill Keith, the author of Louisiana's new law, said he supplied Treen with textbooks and materials to support creationism's scientific claims. One wishes that the governor would have had this evidence corroborated by the National Science Foundation or the American Academy for the Advancement of Science. In doing so, Treen might have learned that creationism is based on faith in the supernatural, not scientific investigation. It is

religion that this new law is all about — fundamentalist religion. As such, it has no place in science courses.

Robert Paul Roth, who is dean at Luther-Northwestern Seminaries in St. Paul, Minnesota, advised the nation's school boards: "Equal time legislation will create bad science, bad theology, and bad pedagogy. Evolution can never threaten the belief in God and creation. And if left alone, scientists will teach all competing theories."

There were many scientists, theologians and educators in Louisiana who understood that creationism lies outside science. Some of them publicly opposed the monkey bill, but most remained silent. Perhaps they were hoping that Dave Treen, unlike his counterpart Frank White in Arkansas, would not subject his state to ridicule by signing a monkey bill into law.

If so, they guessed wrong. The monkey law is now on the state's back, and there's likely to be more foolishness before this is over.

The Times of Shreveport also tried to sway public opinion away from support for creation-science. Capitol Bureau Chief John Hill wrote the following opinion column entitled: "ACLU Suit: Saving Us From Ourselves." It said:

> Baton Rouge — The American Civil Liberties Union may save us from ourselves.
>
> The ACLU has launched legal attacks on so-called "creationism" laws in Arkansas and Louisiana, the only two states whose legislatures have seen fit to mandate the teaching of the divine creation of man alongside Darwin's theory of evolution in public school classrooms.
>
> In both suits, the ACLU maintains the laws unconstitutionally establish religion in public schools.
>
> The Arkansas suit will be heard in federal court in Little Rock beginning Monday. If the court there rules the Arkansas law does unconstitutionally establish religion, it will certainly be a precedent which the New Orleans federal court would probably follow.
>
> In both cases, prominent religious leaders and educational groups have joined the ACLU as plaintiffs.
>
> In the Louisiana suit, filed this past week in New Orleans, the plaintiffs include Bishop Kenneth Shamblin, head of the Methodist Church; the Reverend William Hatcher of Ruston, the General Presbyter of the Presby-

tery of the Pines; the Reverend Phillip Allen of Parkview
Baptist Church, Monroe; the Reverend James J. Stovall,
executive director of the Louisiana Interchurch Conference; the National Association of Biology Teachers, Inc.;
the National Science Supervisors Association; the National Science Teachers Association; and the Rabbinical
Council of New Orleans.

The view of the reasoned religious leaders was
perhaps best expressed by Reverend Stovall, who was
once a minister at Monroe's First Methodist Church:

"We of the church believe that science should be presented in the classroom and that creation should be
taught in the home and the church or synagogue.

"Our belief in God as Creator is not science; it is an
act of faith that should be separated. There is no conflict
between the scientific theory of evolution and our faith in
God as Creator. We do not place any time limitation on
how or how long it has taken God to do His work.

"Science should be left free to determine on empirical
grounds the best theory explaining how our universe and
life came into existence. Faith should be left free to
determine on historical and existential grounds the meaning of our lives on planet earth."

Stovall's is certainly not the only voice of reason on
this politically touchy subject. Several others were raised
during the 1981 regular session as Caddo Parish Senator
Bill Keith pushed his creationism bill through the
Legislature.

But legislators were simply chicken. Several admitted
privately they were philosophically opposed to Keith's
bill, but were afraid of the noisy religious fundamentalists back home.

Those on both sides of the issue will be watching the
Little Rock court action with great interest.

There were several things about Hill's article which
disturbed me and I shared those concerns with him in a letter.

First, he declared the ACLU might "save us from
ourselves." I suggested to him that an extremist organization like the ACLU likely would only help the abortionists,
pornographers, draft dodgers and anarchists.

Second, the article quoted only the evolutionists and

didn't give creationists a chance to voice their point of view.

Third, he made the slanderous remark that "legislators were simply chicken" for voting for the bill and inferred they did so because they were afraid of the "noisy religious fundamentalists" back home.

That was so inaccurate and unfair. Statewide polls revealed that 75 percent of all the people in the state favored the bill. Surely there are not that many fundamentalists in Louisiana. The legislature, having debated the issue for over a year, made a rational decision based on all the facts. But reporters, disappointed because they opposed the bill and lost, continued to slur the legislators who supported and voted for the bill.

The Shreveport Journal, in typical fashion, even tried to harass Attorney General Billy Guste by suggesting his support for the creation-science law would hurt him politically. When everything else fails, that's the way a newspaper goes after a public official.

The newspaper wrote an editorial criticizing the way Guste was handling the defense of the law. What it amounted to was them, the opponents, trying to tell him how to defend the law.

Their editorial, entitled "Creating a Legal Mess," said:

> Since William Guste is now in his third term as the attorney general of Louisiana, he should — by this time — feel sufficiently secure in his position to stick to legal reasoning without being too greatly concerned with political expediency.
>
> Unfortunately, as the attorney general's action over the creationism issue indicates, this is not at all the case.
>
> There was never any doubt that the state's creationism law would end up in the courts. Guste, under those circumstances, would then have had the responsibility and the obligation to offer the best defense possible of the state's law.
>
> But instead of allowing the obvious to occur, the attorney general involved himself in a maneuver that reflects poorly on his judgment.

The day before a suit was filed against the law in a New Orleans federal court, the attorney general went to a federal court in Baton Rouge asking that the law be declared constitutional and that Education Superintendent Kelly Nix implement its provisions.

Nix previously had said that he would not spend state money to purchase creationist-oriented textbooks and institute curriculum changes until the courts had ruled definitively on the law's constitutionality.

Nix's reasonable opinion was not satisfactory to some on the creationist side of the fence; thus the attorney general's ploy. Using a meaningless cover of "let's handle this ourselves without outside interference," Guste went to court.

The purpose of a declaratory judgment is for the court to establish the rights of litigants before any actual damage takes place. Unlike an advisory opinion, however, a declaratory judgment demands that the parties involved have a real controversy out of which an injury or a loss is likely to occur. Federal courts are involved with declaratory judgments but not advisory opinions.

Yet Guste's action seems closer to the latter than the former. What damage is he attempting to forestall? What real adversarial controversy exists between the two state officials?

It should be obvious that the true contest over the creationist law revolves around the suit in New Orleans supported by the American Civil Liberties Union, not the maneuver in Baton Rouge.

The attorney general's game-playing appears to be more a matter of keeping him off the firing line than it is rooted in reason. This idea is further supported by Guste's decision to hire two outside attorneys to handle the state's case. So much for any concern with "outside agitators."

Having witnessed the controversy that the Arkansas attorney general got himself into in that state's court fight over creationism, it seems as if Guste decided to avoid any flak by staying home during the battle. This is simply not the attitude that should be taken by an attorney general.

In his apparent vacillations to avoid a fire fight, Guste also committed the egregious error of saying that the outside attorneys from California and Virginia would be paid from privately raised funds. So now we have the specter of the state's case not only being argued by outsiders, but being paid for by private interests.

It is, to say the least, an entirely bizarre way for the attorney general to go about his business. "Blowing in the wind" may be a nice title for a song, but it is far from a comforting description when applied to the state's top-ranked legal official.

These are just a few examples of how the newspapers fought so hard to kill the creation-science law — and failed.

Louisiana State Senator Dan Richey, a strong proponent of the creation-science law, has made some interesting comments about the news media. Speaking out on how the media reported a certain speech by President Ronald Reagan, the senator said:

"The president defined the problem — government overspending — and then mapped out additional ways to reduce our dependence on the federal government.

"Within minutes after his speech, the national media began an unprecedented attack on his administration. . . .

"If you had the misfortune of watching the president's speech on CBS, you noticed the news reporters tear into the president's program like sharks. They sent out so many negative vibrations in five minutes it's a wonder our TV sets didn't explode. . . .

"CBS and other representatives of the national news media have made a mockery of reporting the news. Their tactics are nothing short of 'yellow journalism,' a phrase coined several generations ago to describe news distortions and exaggerations. . . .

"This may end up having a great positive impact on national politics. . . . Because the sooner the people realize just how biased the national media really is, the less likely they are to be influenced by such media misfits."

Somehow, in the 1970s, people in the media began equating intellectualism with right and the result was that nihilism, the idea that traditional beliefs are unfounded, spread through the journalistic community. A sort of *National Enquirer* syndrome pervaded the journalism profession.

People are reacting to this "anything goes" journalism. The time may soon come when the public will call the media to account for the distortions of the news and bias in news content.

The public should boycott newspapers — and their advertisers — that advertise abortion and pornographic movies; that distort the news; and that print trash and filth.

Yes, in a free society, a newspaper can print whatever it wants to print. But the public also has the right to boycott the newspapers and their advertisers to protest the contents.

The time will come, and maybe soon, when decent, patriotic citizens will take a stand against a media which has prostituted the freedom of the press into a freedom for license and irresponsibility.

U.S. News & World Report published an article in the June 29, 1981, issue entitled: "Uneasy Press Sets Out to Refurbish Its Image." In the story the news-magazine chronicled a number of incidences where newspaper reporters and writers were guilty of telling half-truths, fabricating stories and writing lies about people.

Here are some of the examples:

> The *New York Daily News* discovered that one of its writers, Michael Daly, could not substantiate parts of a column about violence in Northern Ireland. He resigned after admitting that he had used a fictitious name to identify the column's central figure.
>
> *New York Post* columnist James Wechsler charged that a story in the *Village Voice,* a New York City weekly, gave a misleading impression that reporter Theresa Carpenter had interviewed Dennis Sweeney, accused killer of former Representative Allard Lowenstein. The charge was sustained in mid-June by the National News

Council, an independent watchdog of the press. Ironically, the story was among those that won Carpenter the Pulitzer Prize for feature writing that had been returned by the *Washington Post.*

An associate editor of the *Portland Oregonian,* Wayne Thompson, admitted that he fabricated quotes attributed to then-Governor Dixy Lee Ray of Washington. Thompson said that he had interviewed Ray, but that his tape recorder malfunctioned, so he reconstructed the interview from memory and scanty notes. Thompson was suspended for 60 days.

The New York Times Company said that two stories about Polish unrest from free-lancer Harley Lippman, distributed by a *Times* subsidiary, "appear to be of questionable accuracy." The author of the articles, which ran in three newspapers in the U.S. and three abroad, defended them as "truthful in their entirety."

The magazine also said, "The nation's journalists, who for generations have been examining other people's foibles, now find themselves under the public's microscope.

"Widespread charges of fabricated stories, inaccurate reporting and lax editing are forcing publications to re-examine standards and tighten supervision of writers and editors. . . . "

According to the magazine, many editors blame the problem on the "new journalism."

"Exponents of new journalism use composite characters, report events and dialogue that they neither saw nor heard and recount the innermost thoughts of subjects in a combination that can blur the line between fact and fiction," the magazine reported.

With this in mind, it is not surprising that the media — and particularly newspapers — tried to crucify the creationists in Louisiana. But they failed and will continue to fail because the people have a free-born sense of right and wrong. And nothing, including newspapers, can take that away from them.

CHAPTER 7

The ACLU Con Game

The American Civil Liberties Union (ACLU) is the legal arm of Secular Humanism in our society today and the strongest opponent of the balanced treatment of creation-science and evolution-science in the public schools. The organization has become a kind of intellectual Ku Klux Klan.

Let's look carefully at the ACLU so that we may better understand this left-wing, extremist organization which is guilty of meddling in so many issues which are none of its business.

The ACLU is a very powerful, well-funded legal organization which has worked long and hard on behalf of a host of radical causes. Some of them include:

• The right for a woman to have an abortion, thus killing helpless, defenseless, harmless unborn babies.

• The right for people to burn the American flag and draft cards and to dodge the draft.

• The right for pornographers to produce and distribute all forms of pornography.

• The right for Nazis to march in Skokie, Illinois, regardless of how the majority of the townspeople felt about it.

• The right of communists and communist-front organizations to advocate the overthrow of democracy in America.

• The right of Marxist professors to teach Marxism in taxpayer-supported state universities.

• The right of illegal aliens to come to the United States and take jobs away from American workers.

• The total freedom for all kinds of gambling.

Contrawise, the ACLU has vigorously opposed:

• Equal time for creation-science in the public schools.

• Voluntary prayer in the schools and other public bodies.

• Displays of manger scenes at Christmastime in any public building.

• Laws which would strengthen the police to help bring law and order to our society.

• The Congressional Committee on Un-American Activities, the official committee charged with investigating communist acts of conspiracy in this country.

• Laws that would restrict homosexual school teachers.

The ACLU is certainly misnamed for the only liberties they want to protect are those of the abortionists, pornographers, draft dodgers, prostitutes, gamblers, Nazis, communists and flag burners. But they oppose the civil liberties of school children to have all the scientific truth regarding origins. They are also quite strong on defending freedom of speech — their speech, but not yours.

Let's examine the origins of the ACLU.

The Review of the News in its August 13, 1975, issue published an informative and provocative account of the ACLU's background. It was entitled "The ACLU Con Game" and we reproduce portions of it here.

> The ACLU held its 50th Anniversary dinner in Irvington, New Jersey, on October 17, 1970. Honored at this dinner were identified Communist Paul Robeson . . . admitted Communist David Dellinger; and C. Willard Heckel, active defender of Communist Angela Davis.

Each of these men was presented the "50th Anniversary Civil Liberties Award." Identified Communist Pete Seeger presented the award to Paul Robeson.

It is not unusual for the ACLU to honor Communists; after all, it was Communists, Socialists, and Anarchists who founded it to serve their revolutionary cause.

The American League to Limit Armaments was founded on December 18, 1914, as an offshoot of the Emergency Peace Federation, headed by Communist Louis Lochner. Organizers of the League were Jane Addams, a long-time Socialist, later exposed as a secret member of the Communist Party; John Haynes Holmes, another early Socialist leader later active in Communist Party affairs; George Foster Peabody; Stephen Wise, yet another Socialist pooh-bah later active in Communist activities; L. Hollingsworth Wood; and, Morris Hillquit, a founder of the Socialist Party and later a paid agent of the Soviet government.

In 1915 the League changed its name to the American Union Against Militarism and began an intense anti-draft campaign. During this period it opened up a Civil Liberties Bureau to protest draft laws. In October of 1917, an Anarchist/Socialist named Roger Baldwin was the director of the Civil Liberties Bureau and reorganized the group as the National Civil Liberties Bureau. Baldwin was thrown in jail for a year, but resumed his duties in 1919. The next year the NCLB became the American Civil Liberties Union.

The original National Committee of the ACLU contained such worthies as Elizabeth Gurley Flynn and William Z. Foster, both later chairmen of the Communist Party; radical Communist Scott Nearing; Socialist Party chairman Norman Thomas; and, others of a similarly crimson hue. In 1920, Roger Baldwin was Director of the ACLU, Harry F. Ward was Chairman, and Louis Budenz was Publicity Director. Dr. Ward was a long-time professor of Christian ethics at Union Theological Seminary and was exposed as a secret member of the Communist Party. Budenz was a top official of the Communist Party. . . .

It was Roger Baldwin who made the ACLU what it is today. Baldwin ran the show from 1920 until 1950, when he retired as potentate to become chief advisor. What is his philosophy, and what motivated him through his thirty years as boss of the American Civil Liberties Union? The best source for an answer to that is the man himself. In 1935, ACLU Director Baldwin clearly stated his objective as follows: "I am for socialism, disarmament, and ultimately for abolishing the state itself as an instrument of violence and compulsion. I seek the social ownership of property, the abolition of the propertied class and sole control of those who produce wealth. Communism is the goal." There is no evidence that he ever deviated from that purpose.

On the contrary, a look at the history of the ACLU indicates that Roger Baldwin and his followers have consistently used the organization to achieve the goals of Communism in our country. In a cursory check of the ACLU Board of National Committee Members elected since 1920, *American Opinion* magazine found in the reports of investigating Committees of the Congress that almost 80 percent of them had affiliated themselves with Communist activities over the years.

Breaking down their Red Affiliations, of the original ACLU Board of Directors, 36 had affiliated with a total of 672 officially cited Communist Fronts. Of the ACLU officials elected in the 1920s, all but one of the 14 had Red Front citations, with the total amounting to 101. In the 1930s, 30 of the 32 elected firebrands amassed a total of 422 Communist Front citations. Of those chosen as officials in the 1940s, 24 of the 34 had joined 296 Red operations. Those elected in the 1950s racked up a total of at least 52 such citations, while those chosen for top spots in the 1960s had attained at least 67 Communist Front citations prior to 1941. Twenty-one miscellaneous ACLU local officials around the country were also checked by *American Opinion* and found to have amassed a total of at least 204 Communist affiliations — bringing the grand total for 206 leading members of the ACLU to the incredible sum of 1,754 officially cited Communist Front affiliations.

Investigative Committees at both the state and national levels have repeatedly issued official reports warning of the ACLU radicalism. As early as 1920, for example, a Joint Committee of the New York State Legislature took the testimony of many witnesses, including Roger Baldwin, and reported: "The American Civil Liberties

Union . . . in the last analysis is a supporter of all subversive movements; and its propaganda is detrimental to the interests of the state. It attempts not only to protect crime, but to encourage attacks upon our institutions in every form."

A September 1923 report of the United Mine Workers of America warned: "Active among the 'intellectual' classes of the country and posing as a champion of the 'liberties of speech, press, and assembly,' is the American Civil Liberties Union, at New York. This organization is working in harmony and unity with the Communist superstructure in America . . . conducting a nationwide campaign for the liberation of Bolshevik agents and disloyal agitators who have been convicted under the wartime laws or the syndicalist laws of different states for unpatriotic or revolutionary activities."

On January 17, 1931, the Special House Committee to Investigate Communist Activities in the United States reported: "The American Civil Liberties Union is closely affiliated with the Communist movement in the United States and fully 90 percent of its efforts are on behalf of Communists who have come into conflict with the law. It claims to stand for free speech, free press and free assembly but, it is quite apparent that the main function of the ACLU is to attempt to protect the Communists in their advocacy of force and violence to overthrow the government, replacing the American flag by a red flag and erecting a Soviet government in place of the Republican form of Government guaranteed to each state by the Federal Constitution."

A Navy Intelligence report read into the Congressional Record on September 10, 1935, cited the American Civil Liberties Union. "This organization is too well known to need description. The larger part of the work carried on by it and its various branches does undoubtedly materially aid Communist objectives. . . . "

In 1938, the Special House Committee on Un-American Activities reported: "Not only does the American Civil Liberties Union admit its open defense of Communists . . . but it also admits that it has loaned considerable money to the International Labor Defense, a Communist movement, which is a branch of the 'Red' International Aid of Russia. Only a short time ago the Union received a refund for bails which it had furnished for the Bridgman (Michigan) Communists. The trial was abandoned by the government because of the death of the main witness in the case. The Union likewise admits that it has received funds from the (Communist) Garland Fund."

A 1943 report of the California Fact-Finding Committee on Un-American Activities formally concluded: "The American Civil Liberties Union may be definitely classed as a Communist front. . . . At least ninety percent of its efforts are expended on behalf of Communists who come into conflict with the law."

The ACLU has been at the forefront of striking down almost all of our laws which would control abortion in this land. Since 1973, the biological holocaust of abortion has taken the lives of some eight to 10 million little unborn babies. This is even worse than the Nazi holocaust under Adolph Hitler during World War II. The question is, "Why?" Why would the ACLU favor abortion? If they believe in civil liberties, what about the civil liberties of the unborn child?

It is all a part of their original goal. They know that a nation cannot long survive that allows the murder of its unborn infants. So, abortion fits into the master plan of the ACLU and those groups that oppose a democracy where the will of the majority is supposed to prevail in all matters. They hope to see that democratic process replaced by socialism, whereby the state would control the lives of the people.

Why would the ACLU work so feverishly on behalf of draft dodgers and flag burners? Because they follow the secular humanism philosophy which believes that in order to reshape this nation around socialistic principles, patriotism must be destroyed. They know that as long as people love this country, love their flag and believe in traditional morality, they can never witness this nation's downfall.

The ACLU has filed suit in all 50 states of the union to strike down the laws controlling and prohibiting pornography. They use the old, wornout argument that says: "No one should be able to tell me what I can read, write or publish."

What is wrong with that kind of demented reasoning?

For one thing, my police officer friends tell me that some men who hang around pornographic movies often become so stimulated they go out and rape innocent victims or expose themselves to little children on school playgrounds.

So it is not so much a question of freedom for pornographers as it is the right of decent people to protect themselves from the results of that pornography. Yet the ACLU says no majority of people has the right to establish laws to protect themselves and their children from filth of any kind.

What about "kiddy porn"? Should an evil man or woman have the right to force little children to go through all the acts of perversion on camera just to prove that man or woman has certain civil liberties? What about the civil liberties of the children who often suffer irreparable physical and psychological injury? Society has the right and the obligation to protect those children.

The ACLU even defended the American Nazi Party's right to conduct a march through the streets of Skokie, Illinois, over the strong protest of the predominantly Jewish community who remembered the holocaust in Nazi prison camps and the extermination centers in Europe during World War II. Some of the residents of Skokie were concentration-camp survivors who argued that it would be sheer cruelty to them to allow the Nazis to march through their streets. The city government agreed and the ACLU filed suit against them and won. However, the march never materialized.

The Columbia Broadcasting System (CBS) produced an intriguing television special "Skokie" which dealt with the sensitive issue of the Nazi's right to march.

During one very dramatic scene in the program, an elderly Jewish man who had been a supporter of the ACLU, asked the question: "Should we defend the rights of those who, if they came to power, would take away the rights of those who defended them?" For that question, the ACLU lawyers had no answer.

The most serious threat the ACLU poses for those of us who believe in a free society is their support of the right of communists and communist-front organizations to advocate the overthrow of democracy in our land.

Why would they advocate such a treacherous doctrine? Again, is it all a part of their master plan which would allow a minority of socialists to take over this land. That is why the ACLU defends the rights of professors in our taxpayer-supported universities to teach Marxism.

We who believe in democracy should always remember what it is. Democracy is a system based on the will of the majority while providing protection for the rights of the minority. Thus, democracy must protect the minority from all forms of discrimination and help those who make up the minority realize their dreams.

The ACLU believes that pornographers, Marxists, abortionists and communists — who are minorities in this land — should be able to do exactly as they please, regardless of the will of the majority. They confuse democracy with license to do as one pleases. But when that license poses a threat to the democratic society, those who believe in the democratic process have the obligation to oppose it.

Another emphasis of the ACLU always has been to defend the rights of criminals. They have been quite successful. As a result of their legal activities, criminals now have more rights in this land than victims of their crimes.

The ACLU has been particularly successful in crippling the investigative powers of police and getting many criminals released from jail.

Some years ago the New York Police Department planned new ways to deal with the epidemic of crime in the city. Police officials decided they would concentrate on trying to curb the criminal activities of second, third and fourth offenders. The ACLU immediately filed suit against the city, saying that violated those criminals' civil rights.

Barrons, the *National Business and Financial Weekly,* assigned a writer to do an in-depth article on the extremist ACLU. It was entitled "The Curious Story of the ACLU." Here's what *Barrons* said about them.

"Violence and civil liberties these days seem to be inextricably entwined. Chances are that whenever violence erupts, someone representing the American Civil Liberties Union is already on the spot or quickly appears to jealously guard the rights of the violent ones. A famous . . . instance: shortly after TV viewers witnessed the brutal murder of Robert F. Kennedy, an official of the ACLU hurried to protect the civil liberties of the Senator's alleged assassin . . ." *Barrons* reported.

"Careful study of ACLU cases . . . reveals that nearly all the causes it has taken up tend to weaken law and order and the ability of society to defend itself. Some landmark cases give communists more freedom to destroy the nation from within. Those involving the draft erode the state's ability to defend itself against armed attack. Other significant ACLU cases diminish the authority of schools and police and the influence of religion,"*Barrons* wrote.

Barrons also noted the ACLU stand on certain issues is quite curious politically as it relates to the concept of the separation of church and state.

For instance, they favor churches being used for federally funded Head Start programs and even using nuns as teachers. Yet they argue that "one nation under God" violates the concept of the separation of church and state.

The respected *Review of the News* called the ACLU "a radical organization founded by communists, socialists and anarchists to attack our society."

According to the news magazine:

To criticize the American Civil Liberties Union and
its activities is abhorrent to all those noble souls whose
eyes fill up at the very mention of incarcerating the
criminal or depraved. The ACLU has earned a nationwide reputation for its effective activism in defending the
special interests of draft dodgers, murderers, homosexuals, communists, black power groups, and assorted
malcontents making war on our once orderly society.

Throughout its history, the ACLU has been agitating
for political and social change, but its 50th Anniversary
in 1970 seemed to mark an even more pronounced move
toward radical activism. John Pemberton, director of the
ACLU at the time, predicted that his organization would
greatly expand its interests. Future emphasis, said
Pemberton, would be on "bringing in the large classes of
people who've essentially been excluded from civil liberties — the poor, the non-white, youth, prisoners. We need
now to recognize the mass nature of the problems. The
survival of democracy demands it. . . ."

There is no radical issue which has escaped the
ministrations of the ACLU lawyers. They have become
active in recent years, for instance, in pushing for the
"liberation" of students in our schools. In October of
1970 these defenders of "civil liberties" sponsored a
"Student's Rights Conference" for high school students
at Rutgers University. Chief agitator at the conference
was Alan Levine, boss of the New York ACLU's "Student's Rights Project." Levine told the assembled
youngsters: "Oppressive institutions give you no right at
all to say why you go there, how long you go there, and
what you do while you're there." He urged them to demand control of their schools and declared: "Indeed, you
cannot exercise the rights the courts have told you you
have without disrupting the system." The message was
clear: Tear the schools apart to get what you want and
we will see that you are not punished by the law. Ask the
people of South Boston if students are getting the
message.

The ACLU says it is now equally concerned with the
"constitutional rights" of prisoners. Two of the most active lawyers in its National Prisoner Project are Philip
Hirschkop and Herman Schwartz. In the March 1973

issue of the ACLU newspaper *Civil Liberties,* they discussed their work candidly. The goals of the National Prisoner Project are cited as follows: "First, getting people out. Next, protection of prisoner's First Amendment activities. Next, reform of pre-trial facilities." Lawyers Hirschkop and Schwartz believe that prisoners, like students, should be given control of the institutions they inhabit. And what is the objective of these ACLU reformers? As they explain it: "The ultimate goal of the prison reform movement is to end imprisonment as we know it today. But that goal is distant. The Prison Project's intermediate goal, therefore, is to end imprisonment now for as many inmates as possible by reforming sentencing, bail and parole procedures and to make imprisonment as tolerable as possible for others." It is all too clear that the idea is to turn the prisons over to the inmates and put as many convicts as possible back on the streets by wrecking present safeguards.

We should note in passing that ACLU lawyer Philip Hirschkop is listed as a member of the National Lawyers Guild, officially cited as "the foremost legal bulwark of the Communist Party."

Women have also been targeted for "protection" by the ACLU. The headlines of the February 1972 issue of *Civil Liberties* declares: "Priority Program for 1972, The ACLU and Women's Rights." Suzanne Post of the ACLU Board wrote: "A woman's very personality is violated by laws denying her control of her own body." She contended that anti-abortion laws constitute "cruel and unusual punishment" by forcing a woman to carry an unwanted child for nine months. The ACLU quickly set up the Women's Rights Project, and began to fight for the "civil right" of expectant mothers everywhere to kill their babies through abortion. Indicative of ACLU's hypocrisy is the fact that while it condones slaughter of the unborn, it opposes capital punishment. In July of 1973 the ACLU informed its members that it was going to serve as a "nucleus" for those fighting efforts at the state level to circumvent the Supreme Court's abolition of the death penalty. To this end it set up a Death Penalty Project.

Like the Communists, the ACLU is also in favor of gun control. Typical is the position of *The Docket,* official publication of the ACLU in Massachusetts. The issue for

April 1974 reports: "CLUM (Civil Liberties Union Massachusetts), favors all bills that seek to control ownership of guns. When firearms are widely owned, there is a threat to free expression of ideas." Forgive us if the logic of that last sentence escapes us.

Founder Roger Baldwin once stated how he planned to accomplish all of his radical objectives. He said: "We want to, also, look like patriots in everything we do. We want to get a good lot of flags, talk a good deal about the Constitution and what our forefathers wanted to make of this country and to show that we are the fellows that really stand for the spirit of our institutions."

Again, here's what the *Review of the News* wrote about Baldwin's camouflage patriotism:

The ACLU disguise of patriotic concern for America's Constitution and our civil liberties is beginning to wear thin. Many Americans with well developed moral sensibilities fail to see how the cause of freedom is to be served by abolishing the internal security laws of this nation, prohibiting capital punishment, legalizing abortion, destroying anti-obscenity laws, supporting a grape boycott, legalizing marijuana and other drugs, defending flag burning as a symbolic expression of free speech protected by the First Amendment, suing school officials for allegedly violating the Constitution by permitting the display of nativity scenes at Christmas, prohibiting children from praying at school, or supporting the so-called "rights" of sex perverts to hold government jobs, teach school, and serve in the Armed Forces.
It is very hard not to agree with J.B. Matthews, long-time chief investigator for the House Special Committee on Un-American Activities, who observed in January 1955: "In thirty-seven years of history of the Communist movement in the United States, the Communist Party has never been able to do as much for itself as the American Civil Liberties Union has done for it." The game is as insidious as it is simple. The 1966 Report of the Counter-Subversive Committee of the National Conference of Police Associations put it this way: "In our opinion, the ACLU and its brother organizations have

mastered the technique of Joseph Goebbels and practiced by Moscow Communists to the nth degree. 'Tell a lie, make it big, and tell it often enough so that soon everyone will believe it.' "

While freedom-loving, patriotic people who believe in the democratic process hide their heads in the sand, the ACLU is busy fighting against our traditional freedoms.

A leader of the ACLU in Minnesota announced earlier this year that the organization will appeal to the legislators in all 50 states asking them to repeal state laws which forbid sodomy, fornication and adultery between two consenting adults.

The New Orleans Times-Picayune published a story on November 10, 1981, which further chronicles ACLU activities. The story, with a dateline in Providence, Rhode Island, said:

> A federal judge barred Pawtucket on Tuesday from setting up a nativity scene that has been part of a downtown Christmas display for 40 years.
>
> Chief U.S. District Judge Raymond J. Pettine said having the city owned figures set up each year by public employees is a violation of the constitutional separation of church and state.
>
> The nativity scene, almost life-size, is the centerpiece of holiday decorations put up by the city in the privately owned Hodgson Park.
>
> Pettine rejected the city's argument that the scene is a secular symbol like the Christmas trees, carollers and Santa Claus figures that surround it. He said the creche is "a profoundly religious representation of the birth of Christ."
>
> "The unmuted religious message of the creche and its prominent position in the display strongly suggest that Pawtucket has aligned itself with Christianity," Pettine wrote.
>
> His decision came on a lawsuit filed by the state branch of the American Civil Liberties Union in December 1980. Pettine postponed a decision then to ensure the case an unhurried hearing.
>
> Nicholas P. Retsinis, administrative assistant to

Mayor William F. Harty, said city lawyers will contest the decision in the 1st Circuit Court of Appeals in Boston. He said it would be "premature" for him to comment on the details of the appeal.

"The gist of the city's case was that the nativity scene was incidental to the Christmas display, and our argument would still be based on that," Retsinis said.

Steven Brown, executive director of the state ACLU, called the ruling "a great victory for people who are concerned about religious freedom.

"When the government gets involved with religion you often see that religious symbols get demeaned and their religious meaning is minimized," he said. "That happened here with the city's contention that the nativity scene, a central symbol of Christianity, was only an incidental part of the display."

Recently in Harris County, Texas, the ACLU filed a suit insisting that three crosses and a large Star of David be removed from a meditation area in a city park.

Unless the people of America wake up, and soon, the ACLU will take away all those freedoms we have cherished since the beginning of this great nation.

Here are some things you can do:

1. Learn everything you can about the ACLU — what it does, how it works and where it is trying to lead America.

2. Form citizens groups to actually oppose its activities.

3. Establish legal defense funds to provide legal assistance to municipalities, counties and states under attack by this well-funded group.

4. Elect citizens to public office — mayors, city councilmen, district attorneys, legislators, attorney generals, governors and congressmen — who will oppose them and what they are trying to do.

5. Educate the public concerning the ACLU's great con game. When the vast general public grasps the truth about this group it will mark the beginning of the end for the radical organization.

Earlier this year the ACLU stated their intention of filing suit against the Balanced Treatment Act in Louisiana. A reporter from the *Shreveport Journal* called and asked for my reaction to the news.

My reply was that their plans did not surprise me. I said they defended abortionists, communists, homosexual school teachers, pornographers, draft dodgers, flag burners and that they advocated the overthrow of democracy in America. I also said they were amoral, bigoted and un-American.

A few days later the left-wing *Journal* editorialized on my comments, and strongly defended the ACLU while attacking me. The editorial carried a headline which read: "The Latest Whipping Boy" and said:

> The only thing missing from a recent diatribe delivered by state Senator Bill Keith against the American Civil Liberties Union (ACLU) was the waving of a list that allegedly offered the names of those dedicated to un-Americanism.
>
> The attack which labeled the organization and its "sympathizers" as being "a dangerous, treacherous, bigoted, amoral group of people" seemed to harken back to the McCarthy era — an era characterized by unsubstantiated charges, demogoguery and fear. This is one kind of *deja vu* the nation can do without.
>
> While Keith argues that the country's "moral fiber" is threatened by the ACLU and liberal judges, the real threat to America is more closely linked to the extremist politics of despair. In the present, as in the past, extremist politics finds conspiracies everywhere and morality absent, save in those who loudly proclaim their brand of political dogma.
>
> The point is not that the ACLU is above criticism. Rather, it is that the criticism voiced by Keith is an exercise in the kind of red-baiting that has earmarked the most unenlightened and repressive periods in American history. As such, the charges should not go unchallenged.
>
> Though he failed to wave a list, the senator did cite "figures" to show why the ACLU is "subversive" and "a communist front." However, he failed to note that perhaps the most embarrassing incidents in the ACLU's

long history were the semi-loyalty pledges required and
the reporting of radicals carried out during the 1950s.

Many people might not like a number of the specific
cases handled by the ACLU. But those cases cover the
political spectrum and are connected to the United States
Constitution's emphasis on individual liberty. The ACLU
does not view the Bill of Rights as an article of conven-
ience, to be followed only when it fits in with the concerns of a particular group. This is more than can be said
of some of the organization's critics.

Senator Keith's use of the idea of un-Americanism
regarding the ACLU reminds one of that term's definition in William Safire's *Political Dictionary*. Safire, a syn-
dicated columnist not known for his liberalism, writes of
un-Americanism as "not conforming to the ideology or
sharing the same values as the particular American using
the term."

No wonder that *The New Yorker* magazine, comment-
ing on the House Un-American Activities Committee in
1948, called the term "essentially a foolish, bad word,
hardly worth the little hyphen it needs to hold it
together."

People may use the word to question the integrity of
others, but what it usually signifies is the intolerance of
the user.

A study titled *The Politics of Unreason* notes that,
"The road to tyranny is indeed paved with good intentions, as long as the political process is seen only as a
struggle between good and bad intentions."

The ability to accept the legitimacy of one's opponent
lies at the root of political extremism. And it is political
extremism that stands as the enemy of the open society.

Senator Keith, of course, is entitled to his views. But
his broadside against the ACLU is as dangerous as it is
misguided.

The editorial writers at the newspaper have tried for
years to silence my voice through articles such as this.
Their governmental, religious and political viewpoint of
secular humanism coincides exactly with the ACLU. So, in
their typical biased fashion, they defend the ACLU —
whom they worship as heroes — and brand me as
"dangerous" and "misguided."

However, there are still a handful of newspapers who disagree with the ACLU. The respected *Arkansas Democrat,* which endorsed creation-science, wrote a very provocative editorial on the ACLU in the May 31, 1981, edition. It wrote:

> The American Civil Liberties Union and a cloud of co-plaintiffs are taking Arkansas' new Creationism Act to federal court — and Attorney General Steve Clark is preparing to defend it. What a great to-do! Only in America, of all free-world countries, could anyone ask a court, as ACLU is doing, to establish an exclusionary doctrine of world origins for public school.
>
> You'd think that ACLU, the self-styled champion of civil liberties, would think shame to argue for differing constitutional evaluations of ideas — let alone demand that, on the question of world origins, all theories except one should be denied in the classroom.
>
> Outside class, free speech and ideas prevail; inside, only theories supposedly sanitized by science will satisfy ACLU.
>
> If there's any libertarian savor in its suit, only ACLU can taste it. Its repressive philosophy has court standing only because successive Supreme Courts have been hardly less narrowminded in parlaying what was originally only a ban on an established government church in an era of government churches into a ban on almost any government association with religion.
>
> That's proudly called "separation of church and state" in secularist circles, and ACLU is the chief champion of it. The founders no more dreamed of establishing this materialistic negativism than they dreamed of coining the phrase separation of church and state — which is found nowhere in the Constitution.
>
> But ACLU swings the slogan like a club in its effort to drive all religion out of public life — and now it's pounding away at what it calls religion in disguise — never mind that the act forbids any mention of religion in connection with the teaching of creationism and offers the theory only as a subject of comparative study.
>
> The ACLU's local lawyer, Philip Kaplan, has the formidable task of making the act say what it doesn't, and he was trying his best to prepare our minds for a twist

even as he filed his brief. Kaplan publicly twitted "conservative" supporters of the act for trying to "violate the separation of church and state," which, he reminded us tartly, guarantees the freedom of religion we cherish.

Then, in the brief itself, he calls the act an "establishment of religion" and — without any supporting citation from the act itself — declares that the law aims at teaching the Genesis account of world origins, complete with a "Divine Creator."

None of that's in the act, of course — only the provision that "creationism" must be taught as science alongside evolutionary theory wherever the latter is taught. But if ACLU is to get anywhere with its suit, it has to establish the act as a religious sneak.

The parallel would be for the state to declare that evolution is intended to teach atheism — and the act does in fact declare that the exclusionary teaching of evolution amounts to a preemptive course in religion, no other theory of world origins being allowed. . . .

It's this monopoly that Attorney General Steve Clark will assail in arguing that the new law merely provides for an alternative scientific account of world origins, the aim being to give school children a broader grasp of "originology" than they're getting now under the evolutionary monopoly that rules all class discussion.

Kaplan & Co. are smart enough not to argue for the provable truth of evolutionary theory; they just say that it's good science and that creationism isn't — creationism being only sneaking Christian fundamentalism in disguise and hence an affront to both the Constitution and the truth.

They should learn the difference between teaching and preaching — and show some regard for the law's provisions. If, as the law requires, teachers are to be neutral and religiously non-committal in imparting information about the two theories, where's the teaching of an official faith? And if there isn't one, then where's ACLU's supposed respect for free speech and the free circulation of ideas? And where's its faith in the native discernment of school children — in their power to decide which theory has the greater factual weight?

ACLU will have trouble handling the state's right-to-know approach while maintaining its supposed respect for the free-speech clause of the First Amendment. It can't argue that the Constitution labels some free speech good and some bad, or declare that ideas should make their way in the world by selective censorship rather than merit. And ACLU can't argue, as Kaplan does in his brief, for the creationism act's denying academic freedom. Academic freedom insists on freedom of ideas.

Unless Kaplan & Co. can persuade the court that the Creationism Act is a course in Christianity, intended to indoctrinate rather than inform neutrally, all we'll be witnessing is a crude effort to impose classroom censorship — with all that implies for ACLU's supposed dedication to free speech, unfettered opportunity to learn and the bedrock proposition of academic freedom that all ideas have equal right of admission to centers of learning.

Father Alvin J. Deem of the St. Jude Mission, wrote a very thought-provoking letter to the *New Orleans Times-Picayune,* earlier this year about the ACLU and school children's rights to know. In the letter, Father Deem — who could hardly be considered a fundamentalist — said:

> One of the best answers to the frantic attempt to keep kids from learning the "creationism" theory of origins has to be the letter by the high school senior, Lynda Gallien (Jan. 8). She calls the theory "beautiful" and wonders why creationism can't be taught right along with evolution.
>
> It's a fair question. One should think the American Civil Liberties Union would sally forth in phalanxes to defend the kids' right to know what such a widely held theory is all about, as they do for kids' other "right," e.g., to have abortions without the parents knowing, to be spared the embarrassment of having to look at a Christian Christmas crib or listening to the explanation of an Orthodox Jewish Hannukah, to be relieved of the harmful consequences of seeing — or worse, listening to — other kids praying in public schools (supported by Christian and Orthodox Jewish taxes far more than theirs), etc.
>
> And the Lyndas (and Larrys) might also ask: Where are the defenders of academic freedom, for students as well as teachers? Where are the professors, the judges, the editors, the radio and TV commentators, the legislators, the heads of education departments, the

lawyers, all of whom, with a few honorable exceptions, seem to have lost their voices? Among this educated elite are there not more than two or three who know that pagan philosophers (Socrates, Plato, Aristotle), at least two centuries before the coming of Jesus, spoke of the necessity of "an uncaused cause," "a prime mover" and that "every effect (creation) must have a cause"?

Could it not be true to say that the real issue in all this discussing and arguing for the liberals, the secular humanists, the naturalists — is not whether evolutionism or creationism should be taught, but rather through the teaching of which theory can students more easily be shunted from getting to the big question, viz., must we, like the pagan philosophers, postulate the existence of a supreme being (maker, mover), possessing intelligence and free will, to understand the origin of all things?

To ignore the conclusions of pre-Christian, pagan philosophers and to state — as the judge in Arkansas did and some prominent persons who could be expected to know better are doing — that creationism is solely the product of Bible and/or religion, awakens the suspicion that our more highly educated leaders of today have had the same chicanery pulled on them by their professors as they are, willy-nilly, pulling on the kids today: They were never taught any history of philosophy, were probably told it wasn't important. . . .

The modern intellectual elitists, the secular humanists, the naturalists and the timid believers all might do well to think of a bit of history. . . . The highly educated French aristocracy gave birth to the rationalism and naturalism that led to the French revolution. At the high point of their enlightened activity, they put a naked woman on the high altar of Notre Dame cathedral, called her "Reason" and worshipped her in unmentionable orgies.

They gave no thought to the fury of unguided human passions as the guillotine fell and fell and fell — until finally it fell on them. And the ultimate irony — the poetic finish, one might say — was reached when one of the men most responsible for teaching the supremacy of man, the uselessness of God, died — if we can believe those who relate it — trying to eat his own excrement. How stunningly appropriate, how perfectly natural!

Contrast newspaper, in its May-June, 1982, issue, published a remarkable article about an ACLU lawyer which shows that some of their own are beginning to question their tactics. The article was entitled: "ACLU Lawyer Speaks Out on Creationism: Origins and Civil Liberties." It said:

> The best coverage I have seen on this issue was in an article by Mr. Robert F. Smith, a Western Missouri Affiliate of the American Civil Liberties Union, entitled, "Origins and Civil Liberties," (published in the *Creation Social Science and Humanities Quarterly,* Vol. 3, No. 2).
>
> After closely following creationist literature, lectures and debates for the past five years, Mr. Smith made this conclusion: "Based solely on the scientific arguments pro and con, I have been forced to conclude that scientific creationism is not only a viable theory, but that it has achieved parity with (if not superiority over) the normative theory of biological evolution. That this should now be the case is somewhat surprising, particularly in view of what most of us were taught in primary and secondary school."
>
> "Creationists have been scrupulous to adhere to strict discussion of science alone. Not religion! Statements to the contrary are false," this ACLU laywer asserted.
>
> "Contrary to the allegations . . . no creationist professors are seeking to 'require public schools to offer courses and textbooks that support the literal Genesis account of creation.' Nor can it be legitimately suggested that scientific creationists are 'disguising fundamentalist religion in scientific jargon,' or that they are working for some covert 'advancement of sectarian religion,' whatever personal beliefs most of them may have. Scientific creationism *is not religious creationism,"* he explained, noting that "legal scholar Wendell R. Bird points out, 'being consistent with religious views does not make it a religion.' "
>
> Leveling with his readers, Mr. Smith then noted that "in 1925, in Dayton, Tennessee, at a time and place where only religious creationism was legally taught, Clarence Darrow (the ACLU attorney in the 'Scopes Monkey Trial'), thought it bigotry for public schools to teach only one theory of origins." (Parenthesis mine,

S.C.S.) "Would the ACLU be any less bigoted were it to demand that a modern-day John T. Scopes be allowed to teach 'only' neo-Darwinian evolutionary theory?" he asks.

After documenting various legal and scientific points, he emphasized that "Religious creationism is rightly to be excluded from any public school system. . . . However, the leading professional organizations of educators and scientists pushing scientific creationism . . . have never sought to introduce religious creationism into the public schools."

Mr. Smith then challenged anyone with "suggestions of ulterior motives on the part of creationists," stating emphatically that "scientific creationism is not the Biblical view of creation in disguise. . . ." Anyone making such a suggestion ought to be prepared to give a rigorous, factual demonstration of it.

I admire the ACLU attorney not just because he has obviously done his homework and knows what he is talking about, but mostly because he has set aside his own personal beliefs and prejudices in a rare display of objectivity as is evidenced in these remarkable closing statements.

"I am not an evangelical, fundamentalist Christian, and my fundamentalist friends do not consider me to be a 'Christian.' I have no ulterior motive for the above statements . . . unless it is the simple-minded notion that the ACLU should maintain the highest possible standards in the protection and expansion of civil liberties. . . . I can see no reason why the ACLU would in any way desire to restrict the entry of scientific creationism into the public school system — whether such entry be provided by mandate or through 'laissez faire' means."

At the moment, creationists seem to be getting a bad rap. It's quite similar to the difficulties evolutionists had in Darwin's day with the discrimination, prejudice, distorted press coverage, et al. Actually Mr. Smith of the ACLU summed it up when he accepted Mr. Darrow's judgment that it would be "bigotry for public schools to teach only one theory of origins," and according to that judgment it seems we have a few bigots left today, doesn't it?

How long will the American people allow the ACLU con game to continue?

CHAPTER 8

Disaster in Arkansas

On Tuesday, March 17, 1981, the Arkansas House of Representatives passed a Balanced Treatment Act by a 69 to 18 majority. The measure, which had already won approval by the Senate, was sent to Governor Frank White who had earlier said he would gladly sign it into law.

Thus, Arkansas became the first state in the nation to require by legislative act that creation-science be given equal time with evolution-science in the public school classrooms.

Arkansas members of the House who were proponents of the law cheered when they saw the record of the vote.

Senator Jim Holsted, a democrat from North Little Rock who authored the bill, was jubilant following the great victory. He hailed the bill's success during a news conference. He reminded the news media that his bill limited instruction on the subject of origins only to "scientific evidences" and that it prohibited any religious instruction in the classroom.

"Evolution is the only theory being taught (in the classroom), so students tend to take it as a fact," Senator Holsted told reporters. He added that any course of instruction dealing with the origin of man would now be required to give both theories.

Senator Holsted's Balanced Treatment Act shocked evolutionists who had so strongly opposed the concept. Their front-line bastion — evolution in the schools — had received a crushing blow, not fatal, but serious.

When the cheering stopped, the creationists knew they faced an uphill battle in defense of the courageous new law.

The ACLU immediately said they would challenge the law in court, and they did on May 27, 1981, when they launched a frontal assault on the new law hoping to exclude creation-science from the public schools of the state.

Their complaint which was filed in federal court in Little Rock, said the Balanced Treatment Act:

• Violates the First and Fourteenth Amendments to the Constitution.

• Constitutes an establishment of religion and abridges the academic freedom of both teachers and students.

• Is impermissibly vague.

The ACLU's own complaint could have been viewed as a positive statement for balanced treatment.

It said they were not asserting "the final truth of any theory of evolution."

If it is not "final truth" why did they file suit against another concept of origins being presented to children in the schools?

The complaint said many of the plaintiffs are not "anti-religious" and stressed that many of them believe religion is important in personal life, family life and in the community.

"All plaintiffs are united in the firm conviction that religion is strengthened by its complete separation from government, and that government supported education in science is strengthened by its complete separation from religious doctrine."

The plaintiffs included:

• Reverend Bill McLean, the principal officer of the Presbyterian Church in Arkansas.

• Bishop Ken Hicks, of the United Methodist Church of Arkansas.

• The Right Reverend Herbert Donovan, Bishop of the Episcopal Diocese of Arkansas.

• The Most Reverend Andrew J. McDonals, Bishop of

the Catholic Diocese of Little Rock.

• Bishop Frederick C. James of the African Methodist Episcopal Church of Arkansas.

• Reverend Nathan Porter, a Southern Baptist minister.

• The American Jewish Congress, with members in Arkansas.

• The Union of American Hebrew Congregations, with members in Arkansas.

• The Arkansas Education Association.

• The National Association of Biology Teachers.

• The American Jewish Committee, with members in Arkansas.

• The National Coalition for Public Education and Religious Liberty (National PEARL), with members in Arkansas.

• Other Methodist pastors, teachers, parents, a professor and an attorney.

That array of clerical heavies would be enough by itself to scare the devil out of the creationists! When I read that sterling array of men of the cloth who opposed creation-science, I asked myself: "Do they really believe the Creator formed the earth yet left no evidence whatsoever of his creative act?" Perhaps they think he hid the evidence of creation or created *in cognito* or in secret.

The ACLU complaint continued:

"The Creationism Act is the latest attempt in a long-standing pattern and practice of the State of Arkansas to promote religion in its public schools, to establish a particular religious dogma, and to disparage science when it is deemed to conflict with or be antagonistic to that religious dogma."

Those who carefully read the ACLU complaint saw that it was filled with half-truth and innuendo.

Through the years the people of Arkansas had tried to get the lie of evolution out of their public schools. In 1929 the legislature passed a law which provided criminal penalties for any public school or university "to teach the theory or doctrine that mankind ascended or descended from a lower order of animals." They even went so far as to make it "unlawful and criminal" for anyone in the state to select textbooks or materials which even mentioned the theory of evolution.

However, in 1968 the U.S. Supreme Court decreed that the anti-evolution law in Arkansas was "an attempt to blot out a particular theory" because it conflicted with the concept of creation and therefore constituted the establishment of religion.

The ACLU said the words "creation" and "creation-science" constitute religious doctrine and "embody and reflect particular religious beliefs not shared by adherents of other religious beliefs, or by those who hold no religious beliefs."

The complaint made several other charges about the Balanced Treatment Act:

• "Creation" as used in the act, necessarily encompasses the concept of a supernatural Creator.

• The concept of a supernatural Creator is itself an inherently religious belief and "creation-science" cannot be taught without that belief.

• "Creation-science" is not science. Any scientific statement must be subject to disproof. Because creationism concludes that there is a supernatural Creator or supernatural process, it is not subject to disproof.

• It requires teachers to teach as science a doctrine which they, as professionals, believe has no scientific basis or merit.

• It prohibits the teachers from expressing, and the students from learning, the teachers' professional opinion concerning the relative scientific strengths or weaknesses of either.

• A lot of teachers — if required to give equal time — will teach neither, thereby depriving their students of the right to acquire useful knowledge.

• The Balanced Treatment Act is vague, inconsistent and does not explain to teachers what can and cannot be taught.

When the ACLU accompanies Methodist, Presbyterian and Southern Baptist preachers to court they wear a velvet glove. But when they defend abortionists, pornographers, homosexual teachers, communists and anarchists, they display an iron fist. That is the character of this extremist organization.

The suit was filed and things settled down in Arkansas for a time.

But the creationists sensed an ill wind blowing across the state. Several incidents caused them concern. They included the following:

1. Attorney General Steve Clark, constitutionally required to defend the law, publicly said he had "qualms" about balanced treatment.

2. The ACLU began pouring New York lawyers into Arkansas to fight the law, while Clark appeared to be making little preparation for its defense. Look at the facts:

• Six weeks prior to the opening of the trial, Clark had assigned only one assistant attorney general to the case, a major case of national scope which should have had a battery of a dozen of the best constitutional lawyers available to try the case for the state.

• One New York law firm volunteered 14 of its lawyers to work fulltime, as needed, to assist the ACLU in Arkansas. They were allowed to charter jets to carry them all over America to take depositions, interview witnesses and conduct research related to the case.

• Federal Judge William Overton, scheduled to hear the case, denied a request by a group of creationists to intervene on behalf of the law. The creationists realized that unless someone began preparing an adequate defense, the law would never have a chance of standing up in court.

• Only three weeks prior to the beginning of the trial, Attorney General Clark participated in a fund-raising event on behalf of the ACLU. They were raising money to fight the Balanced Treatment Act which Clark was to defend. His act appeared to creationists to be a serious conflict of interest and caused great concern among the ranks of creationists in the state.

The defeat of the law was a foregone conclusion after Judge Overton ruled that outside parties could not intervene.

Why were they so crucial to the case? Because great expertise is required to defend creation-science. And there was no one in the attorney general's office who had that expertise.

Also, those familiar with creation-science realized defending the Arkansas law would be a gargantuan task.

Attorney General Clark didn't fully realize the scope of the trial and later so admitted. But it was too late.

The applicants for intervention on behalf of the state included a formidable group who supported creation-science. They consisted of taxpayers, citizens, science professors, religious persons (Jewish, Muslim, Catholic, Mormon and Agnostic), public school teachers and parents.

Some of them were:

• Arkansas Citizens for Balanced Treatment in Origins, an organization of science professors and other science professionals.

• Committee for Openness in Science, an organization of more than 600 individuals nationwide who hold doctorate or masters degrees in science and support balanced treatment.

• William C. Gran, a physics instructor at the University of Arkansas in Little Rock.

• John M. Denton, instructor of chemistry at Westmark Community College.

• Dr. Lawrence A. Meek, associate professor of physics at Arkansas State University.

• Dr. Robert L. Bond, professor of instrumentation at

the University of Arkansas.
* Dr. Daniel B. McCollum, associate professor of mathematics at the University of Arkansas.
* Dr. Ralph Fell, who holds a Ph.D. in biochemistry and is engaged in biochemical research.
* Dr. W.R. Collie, former assistant professor of pediatric genetics at Arkansas School of Medicine in Little Rock and now a physician.
* Rabbinical Alliance of America.
* Dr. Asadollah Hayatdavoudi, a Muslim, chairman of the Department of Petroleum Engineering at the University of Southwest Louisiana.
* Irving D. Saeger, chairman of the Science Department and a biology teacher at Parkview High School in Little Rock.
* Arkansas Association of Professional Educators, comprising many public school teachers.
* Dr. W. Scott Morrow, a science professor and agnostic who, though an agnostic and evolutionist, believes there is a vast amount of pure scientific evidence which supports creation-science.

The motion to intervene was filed by attorneys Wendell Bird, a Yale Law School graduate of San Diego, California, and John Whitehead of Manassas, Virginia. Both men are experts in the areas of creation-science and First Amendment freedoms. They would have been of invaluable assistance to Attorney General Clark had they been allowed to work with him.

However, after Judge Overton denied the motion to intervene, Clark also refused their voluntary offer to help.

Both men had researched the balanced treatment concept for years and were thoroughly versed on the issues. They also had written scholarly articles for law journals — including Harvard and Yale — on the subjects.

Most of the intervenors were scientists while most of those joining the ACLU opposing the law were liberal bishops, pastors, ministers and rabbis.

The motion to intervene clearly established a basic defense strategy. It said:

• The Balanced Treatment Act conforms to the requirements of the establishment clause of the First and Fourteenth Amendments by bringing about the required neutrality of public schools toward religion.

• It has the purpose and effect of teaching all the scientific evidences about the origin of the world and life, rather than indoctrinating students only in the evolution-science explanation and censoring evidence that supports the other view.

• No excessive governmental entanglement results and no religious or political divisiveness occurs from balanced treatment that does not already exist from teaching only evolution-science.

• The act implements and fulfills the academic freedom and freedom of speech guarantees of the First and Fourteenth Amendments.

• It gives students the academic freedom to hear and to receive information about origins and to assess alternative scientific explanations of origins.

The people who wanted to intervene — to help the state — openly stated in the motion that they did not believe their interests would be fully represented in the trial. They noted that some of the defendants such as the Arkansas Board of Education (which had been sued by the ACLU) opposed the Balanced Treatment Act and wouldn't be expected to vigorously defend it in court.

Dr. W. Scott Morrow, the science professor-evolutionist-agnostic, supported the motion by creationists to intervene in the case. He is an associate professor of chemistry at Wofford College in Spartanburg, South Carolina, and formerly was associate professor of biology at Concord College in West Virginia.

Why would an evolutionist and agnostic be in favor of equal time for creation-science in the public schools? Because he is honest enough to admit there are valid scientific evidences which support creation and he is not threatened or intimidated by those evidences as most evolutionists are.

"I am an evolutionist, and I believe strongly in public schools teaching both creation-science and evolution-science," he said in an affidavit in support of the motion to intervene. "I personally believe that evolution-science possesses more experimental strengths, and in the origin of life area I believe that the initial life forms evolved by the various mechanisms. . . . "

On the other hand, Dr. Morrow acknowledged certain defects in evolution-science and said both its strengths and weaknesses should be presented to public school children.

"I prefer confrontation of the evidence and argument for and against facts or concepts, because through this process one is more likely to derive what could be called the truth or at least a useful working explanation, and it creates intellectual flexibility, develops analytic capacities, and increases student interest," he said. "Although I do not personally believe that creation-science is the better explanation, creation-science brings a proper amount of reservation or questioning to the evolution-science explanation and also presents a plausible alternative.

"Creation-science writings point out many real difficulties or defects with evolution-science. . . . Creation-science also involves positive scientific evidences that should be presented and assessed. . . .

"There is no particular reason why a person has to be religious in a normal sense of the word to believe in creation-science or to believe that it should be given balanced treatment. I see no difficulty in accepting a creation-science explanation for the origin of living things without believing at the same time that the creator was a personal deity.

"Both creation-science and evolution-science can be equivalently scientific, just as they can be equivalently religious. I believe the best way to find truth is not to talk dogmatically about one conceptual framework but to consider pros and cons and to assess alternative conceptual frameworks."

The creationists simply believed they would not be adequately represented in the trial.

But Judge Overton was not convinced and ruled to exclude them from the case.

Thus, prospects were very dim for winning in court.

There were a lot of behind-the-scene facts concerning Attorney General Clark's lack of adequate preparation to defend the law.

Here are some of those pertinent facts that explain why creationists lost in Arkansas:

• The ACLU filed their suit in May of 1981. Clark asked for and received an unusual two-month extension for the trial date.

• Although the trial date was originally set for September, Clark and his staff did very little to prepare the anwer to the ACLU's charges.

• Clark vigorously opposed the motion by creationists to intervene in the case. He said the intervenors represented "special interest groups" who would open the floodgates to innumerable interventions.

• His opposition to the intervenors was the only significant action Clark took during the first four months following the filing of the lawsuit.

• Attorney Wendell Bird explained to Clark that this was a major case and would require a vast amount of discovery work. He also told the attorney general that attorneys defending the law must have expertise in four areas: law, science, religion and education.

• During a previous anti-trust case, Clark retained a private legal firm of experts in the field and paid them $750,000 for their help.

• Eyewitnesses estimated Clark and his staff devoted only 20 percent of the time necessary to the defense of the law and less than 10 percent of the amount of time the ACLU lawyers were working to kill the law.

• A legal defense fund in Arkansas offered to pay for the outside legal experts to assist Clark. The offer was rejected.

• During a pre-trial conference, the judge asked for an example of the evidence for creation. The deputy attorney general could not give one.

• The judge also asked whether creation-science is not inherently religious because it assumes a creator. The deputy denied that it involves a creator — "only a force." The deputy backed himself into an untenable position because of his lack of preparedness. He could have based his answer on the unquestionable constitutionality of mentioning a creator in the public schools by referring to the precedent-setting cases involving the pledge to the flag.

• The attorney general failed to make several legal arguments: For instance, he never argued that evolution is a doctrine of numerous religions and that evolution-science presupposes that there is no creator.

Creationists never had a chance in Arkansas.

CHAPTER 9

Creation on Trial

On December 7, 1941, forces of the Japanese Imperial Navy launched a sneak attack on Pearl Harbor. Forty years later — on December 7, 1981 — the ACLU launched its own sneak attack on the Balanced Treatment Act in Judge William Overton's U.S. District Court in Little Rock, Arkansas.

The Overton courtroom was packed with attorneys, witnesses, spectators and dozens of news reporters from all over the world. The judge pounded the gavel and nine days of grueling testimony began.

Much of the following detail of the testimony in the trial is based on the eyewitness account of Cal Beisner, brilliant young Arkansas writer, who covered the trial.

Attorneys Robert Cearley, Jr., of the ACLU, and Steve Clark, for the defense, made their opening arguments.

Cearley told the judge that teaching creation-science in the classroom is a clever and dangerous violation of the First Amendment.

However, Clark said the balanced treatment of creation-science and evolution-science would broaden the school children's concept of origins.

Cearley said the Balanced Treatment Act was an "Unprecedented attempt by the legislature to arrogate to itself to define what science is and to force religion into the schools in the guise of science." He added that creation-science is a sectarian view based on the belief that the Bible is true and said the law "has the clear effect of advancing a literal interpretation of Genesis in the classroom."

Clark countered by saying the act would expand the knowledge they would be receiving on the subject of origins.

According to Cearley, the law was so vague that teachers would not understand it and therefore could not teach both concepts of the origin of life.

However, Clark reacted by saying teachers should be able to use their professional judgment in teaching the two models.

The attorney general also challenged Cearley's contention that creation-science is religion.

"The mere coincidence between the law and some religious beliefs does not entangle the state in religion," he said.

After the opening arguments, the ACLU called its first witness in opposition to the law — Bishop Kenneth W. Hicks of the United Methodist Church of Arkansas.

Bishop Hicks said he opposed the law because he felt it was based on "a literalistic view of the book of Genesis."

"The words 'In the beginning, God created,' I hold very dearly," the Methodist bishop said. "From that point on, I feel it belittles God and does injustice to both religion and science to try to circumscribe the way he did it."

Bishop Hicks said in his opinion the law "seems to be an intrusion of the First Amendment and a mix of church and state. . . ."

Judge Overton — himself a Methodist — listened intently to his own bishop's remarks.

Another witness, Bruce Vawter, the chairman of the Department of Religious Studies at DePaul University in Chicago, also testified against the law.

"This act in its description of what it calls creation-science has as its unmentioned reference book the first 11 chapters of Genesis," he said.

George Marsden, professor of history at Calvin College in Grand Rapids, Michigan, was another evolutionist to speak against the law. He read some excerpts from creation-science literature and quoted Dr. Henry Morris of the Institute for Creation Research as saying that evolution is the foundation of rebellion against the Creator. He inferred that such a statement by Morris meant that creation-science was pure religion.

Attorneys for the state objected to the statement by saying that the Arkansas law prohibited any religious instruction in the classroom and required that only scientific evidence be presented.

But Judge Overton overruled them.

"They can't wear two hats about this," the judge said. "I don't think the writers can call it religion for one purpose and science for another."

During cross-examination, Clark said creation-science could be taught in a purely secular, unreligious manner.

However, Langdon Gilkey, professor of theology at Chicago University, said the word "Creator" cannot be separated from the word "God."

"The idea of a creator, particularly creation out of nothing, has its source in religious tradition," Gilkey said. "I find it unquestionably a statement of religion."

Dr. Michael Ruse, a member of the Department of Philosophy at the University of Guelth, Ontario, Canada, based his testimony on the history and philosophy of science as it relates to evolution and creation. He insisted that true science would preclude calling creation-science a science because it relies on the supernatural.

During the next four days the ACLU called a battery of witnesses who hammered away at the Balanced Treatment Act.

Dr. Brent Dalrymple, assistant chief of the U.S. Geological Survey, said geological evidence shows the earth to be about 4.5 billion years old. He said radiometric dating systems, based on the rate of decay of certain radioactive materials, have proven to be quite accurate. He added that it is unscientific for the creation-scientists to claim the earth is much younger.

According to Dalrymple, he had read numerous creation-science publications.

"Every piece of creationist literature I have looked into so far has had very, very serious flaws," he observed.

The ACLU also called Dr. Harold Morowitz to testify. He is a professor of biophysics and molecular chemistry at Yale University.

Morowitz said "sudden creation" is "outside the realm of science."

The Yale professor argued that creationists insist that the complexity of living things indicates they could not have come about by chance. Creationists move from complexity to improbability, he said. But "the fact of the matter is we do not know the ways in which life comes about."

During cross-examination, Morowitz was asked to calculate the odds for life creating itself. He said the chance that elements would combine in such a way as to create life are 10 followed by one billion zeroes.

Defense attorney David Williams, who assisted Clark in the trial, asked if scientists had ever created life by a flow of energy through various elements. The professor answered no, that all attempts by scientists to create life had proven unsuccessful.

Morowitz stated that creationists misuse the second law of thermodynamics in their arguments that life could not have created itself. He added that certain scientists ignore the fact that the earth is not a closed system and receives energy from the sun, making it an open system and thus the origin of life is possible.

Asked by Williams if the universe itself is a closed system and subject to laws of thermodynamics, Morowitz replied, "I believe there are astrophysicists who believe the universe is an open system and there are astrophysicists who believe the universe is a closed system."

This was quite a crucial point for the defense. Morowitz admitted that some astrophysicists believe, as do creation scientists, that in a closed system, all matter tends toward randomness.

Next the plaintiffs called Stephen Jay Gould, a professor of geology at Harvard University. Gould, an avowed Marxist, also is the leading spokesman for the concept of "punctuated equilibria" or the idea that evolution took place through rather sudden "quantum leaps." Gould believes that is the reason why there are no "missing links" or intermediate forms in the fossil record.

"Gould said that on the basis of a flood model of geology, one would conclude that all forms of life were alive simultaneously, at least before the flood," Beisner wrote. "However," he said, "the fossil record preserved in the strata of sedimentary rocks in the earth shows that the animals were not mixed together, but are 'rather well ordered in sequence of strata, from the old to the new,' and this, he said, showed the flood theory to be wrong."

When asked if he thought creation-science is scientific, he replied, "Certainly not, because it calls upon the intervention of a creator. . . . "

The defense attorney then asked Gould if the evolutionary theory presupposed the absence of a creator. He replied, "No, evolutionary theory functions either with or without a creator, so long as the creator works by natural laws."

Gould admitted under cross-examination that science was not his only reason for opposing the Balanced Treatment Act. He said his own political liberalism was also a motive for his opposition.

Gould said that it is the best judgment of science and the scientific community that "life arose naturally." But he then conceded that the evolutionary concept was subject to question and being proved wrong, just like anything else in science.

The ACLU then called Dr. Dennis R. Glasgow, supervisor of science education in the Little Rock schools. He said teaching both creation-science and evolution-science would not allow teachers to give their professional opinions on the subject, would "damage the security of students" and "lower the student's opinion of the teacher."

Glasgow also charged that including creation-science in the curricula would be damaging to education and would hinder the hiring of good teachers becaue they would not want to teach where they would be required to present something like creation-science which is unscientific.

Under cross-examination, Glasgow admitted that some studies show that students increased in cognitive development and critical ability when both models were presented.

"He said students should be free to question various ideas, but they had not yet developed the capability to judge between views well before they were old enough to leave the public school systems," Beisner wrote. "He also said that his belief that it would be impossible for teachers and administrators to devise a curriculum for balanced treatment presupposed his belief that creation-science is not science but religion."

The plaintiffs also called Ronald W. Coward, a teacher of biology and psychology in the Pulaski County School District in the Little Rock area.

Coward said he personally decides what is good science and what is not and that creation-science is not real science.

He said the Balanced Treatment Act would require changes in the way psychology is taught in the public schools. He explained that psychology presupposes a relationship between humans and various other animals. Some experiments in psychology, to learn about human behavior, would have to be discontinued.

According to Coward, evolution is the key to science and biology and without it his students would be unprepared for college. But he said if he were required to give equal time to creation science he would teach neither.

"He said academic freedom for students meant their right to pursue available information in a field, but that his responsibility as a teacher was to sort out and select what information was legitimate for the students to examine," Beisner wrote.

The ACLU lawyers also called Marianne Wilson as a witness. She is the science coordinator for the Pulaski County Special School District.

She said she had been involved in trying to develop a two-model curriculum and that those attempts failed because no evidence for creation-science could be found.

Asked if she had tried to use the two-model approach as presented in Dr. Richard Bliss' book, she said she had read the book but "threw it in the trash."

"Under cross-examination by Clark, Wilson said she did not believe the state has a right, through legislation, to prescribe curriculum for its schools," Beisner wrote. "She also said that a belief in a recent origin of man and the earth need not be religious. She acknowledged that although four texts now used in the public schools in her county mention creation science, she had not contacted the publishers of those texts to see if they could provide further information on it."

Clark then asked her about scientific evidences and if she were familiar with the work of Dr. Robert Gentry on the subject of polonium radiohaloes. She said she had seen the work but had also thrown it out.

Asked if she threw it out because she couldn't under-stand the work, she replied, "No."

Clark then read from a pre-trial deposition her statement that she had been unable to understand Dr. Gentry's studies on polonium radiohaloes.

The plaintiffs final witness was Dr. William V. Mayer, professor of biology at the University of Colorado, and director of Biological Sciences Curriculum Study, Boulder, Colorado.

"Mayer said since the inception of the BSCS in 1960, he was sure that their decision to make more open reference to evolution in the textbooks they produced would 'wave a red flag before certain fundamentalists,' and lead to conflict," Beisner reported. "However, they had decided that evolution was so central to understanding all of science that for the sake of quality textbooks in the sciences they would have to discuss it in more depth. . . . "

Mayer admitted that evolution does not include the study of the origin of life but that various concepts of the subject are presented. He said those concepts include *pan spermia* or that the universe is filled with the "seeds" of life; spontaneous generation, that life came about through the random chance combination of elements; and the "steady state theory" which holds that life and the universe have existed forever.

"Mayer said the effect on students of teaching creationism would be confusion and division in the classes causing more problems than it would solve by mixing theology and science in a way that 'damages both and is helpful to neither,' " Beisner wrote.

Mayer testified that the belief in creation-science was based on supernatural, divine revelation which is reflected in creation literature which often refers to the Bible and the Word of God. He said religion is the "unifying theme" in creation writings, not science.

Attorney General Clark cross-examined Mayer and referred to a previous statement by the scientist that he could find no reputable textbook which dealt with creation-science. Clark then called his attention to *The World of Biology,* a popular textbook published by Davis and Solomon. He asked Mayer to read a section which listed six of the arguments creation scientists use.

"Clark then asked Mayer which of those he would understand as religious instead of scientific arguments, and Mayer named none of them," Beisner reported. "Clark then asked which were historical references instead of scientific arguments, and again Mayer acknowledged that none of them were historical. He agreed with Clark that none of the arguments was presented in that textbook as any more historical or religious than a book which would present arguments for evolution.''

Clark also asked Mayer if the idea of the evolution of life from non-life could be analyzed statistically. The scientist said it could not.

Mayer said he believed life on earth had a beginning but could not explain it and that no one had been able to create life in a laboratory.

"Clark, referring to Mayer's pre-trial deposition, asked if Mayer had said it 'may well be that creationism is correct about origins,' and Mayer said he had said that, but that he had added that 'even if it were correct, it's not scientific,' '' Beisner wrote.

"Clark then asked Mayer if he believed students had a right to examine controversies, to see several sides of controversial issues, and to take positions on controversial issues without fear of discrimination from their teachers, and Mayer answered all three questions, 'Yes.' ''

One thing Clark did not pounce upon was that Mayer had a conflict of interest in testifying against creation-science and for evolution-science. The Biological Sciences Curriculum Study, of which Mayer is a director, is the publisher of four evolutionary biology books used by many of the public schools in this country.

The ACLU rested its case and the trial shifted gears and the defense called its first witnesses.

Dr. Norman Geisler, professor of theology at Dallas Theological Seminary, was the first witness to testify for the defense.

He said one of the key issues in the trial was a definition of religion. He said that "humanistic" religion centers on mankind and that the religious commitment of Thomas and Julian Huxley, both strong evolutionists, was an example of that religion. He then quoted statements in the *Humanist Manifesto* which make it very clear that evolutionism is a cardinal doctrine of humanism. He also noted that the manifestoes declare that humanism is a religion.

"Geisler said the first line of the preface to a combined publication of *Humanist Manifesto I and II* describes humanism as a 'philosophical, moral, and religious view,' and said that the word 'religion' is used 28 times in the documents, mostly as a description of humanism itself," Beisner reported. "The publication also refers to humanism as a 'quest for transcendent value,' and a 'commitment to abiding values,' and later in the documents it says it is 'necessary to establish' such a religion."

Geisler then produced an article for the court written by Julian Huxley entitled *The Coming New Religion of Humanism* in which Huxley said evolution is critical to humanism and "its gods are made by men."

He also pointed out that Charles Darwin once said that "natural selection is my Diety."

"To show that belief in a supreme being or creator is not necessarily religion, Geisler referred to the teaching of the Greek philosophers Aristotle and Plato," Beisner wrote. "Aristotle believed there was a 'first cause' or 'unmoved mover,' but did not worship or commit himself to that being, and did not posit any moral attributes to it. Plato believed in a 'demiurgos' which served as a creator in his philosophical system, but again did not think of the being in a religious way. . . . "

Geisler said that the belief that there is a creator has no religious significance but belief in a creator does. He added that the belief that there is such a thing as biological evolution is not religious but belief in that concept is.

Dr. Larry Ray Parker, professor of curriculum and instruction at Georgia State University, was another witness called by the defense. He is a specialist in curriculum principles and serves as a consultant to public schools in the development of curriculum.

Parker made it very clear to the court that he believed the presentation of both models of origins is a sound educational principle. He added that it is good for students to see both sides of the question and to allow them to examine the arguments for themselves. He said that students learn better when they have the opportunity to study contrasting viewpoints.

According to Parker, the Balanced Treatment Act would provide for classroom discussions that would be thought-provoking as the students analyzed origins. He suggested that the classrooms should be places where students are taught to think, not told what to think.

When asked to define "balanced treatment," Parker said it normally means the treatment of all materials from an unbiased viewpoint.

When asked if he thought the balanced treatment of creation and evolution violates the concept of academic freedom, Parker replied that the primary function of academic freedom is to allow the students to be exposed to various materials relating to any subject.

Parker also said that teaching only one view on any subject is "tantamount to indoctrination."

Dr. W. Scott Morrow, professor of biochemistry at Wofford College, South Carolina, was also called by the defense to testify on behalf of creation-science.

Morrow said that evolution doesn't even fit into the definition of science and that the primary reason most scientists believe in it is because they "wanted to believe in it, they looked at evidence and saw it one way, and didn't consider alternatives."

According to Morrow, discoveries made by creation scientists receive the "silent treatment" and are not published in scientific journals because they raise questions about evolution. He said the rejection of creationist works could be compared to the earlier rejection of the theory of continental shift, the work of the earliest molecular biologists and others.

Morrow, an evolutionist and agnostic, said both creation and evolution should be presented in the classroom and that both can be taught as scientific or unscientific, religious or non-religious. He added that the teacher who doesn't like creation-science has no excuse for not teaching it.

Morrow said he had studied creation-science and had come to understand how creationists might understand scientific data differently from evolutionists. He also said the schools should be responsive to the citizens who believe overwhelmingly in creation.

"Asked why distinguished scientists who witnessed for the ACLU had said that evidences used by creationists simply are not science, Morrow said, 'They're wrong. They're lousy in their interpretations of this case,'" Beisner reported. "He added that the fact that 'heavyweights favor evolution is of no great matter.'"

He charged that some of the ACLU witnesses such as Stephen Gould, Harold Morowitz and others had ridiculed creation-science because they "don't like it's conclusion" and don't want it to be considered because their minds are closed on the matter.

Morrow asserted that the mathematical probability for evolution — of random formation of life from non-life — is negligible. He said it would be like trying to find a single grain of sand in all the deserts of the earth.

When asked to give some specific facts for creation-science, Morrow listed the following: the statistical impossibility of life evolving from non-life; the absence of transitions or "missing links" in the fossil record; the insufficiency of mutation to bring about large changes in populations; and new discoveries in measuring the age of the earth.

Another witness for the defense was Jim Townley, a chemistry teacher in the Southside High School in Fort Smith, Arkansas.

Townley said that as a chemistry and physics teacher he would like to teach his students something about the mathematical impossibility of life generating itself. He also said students should be given the data which reveals that the earth is not as old as the evolutionists say it is.

"However, Townley said, he had not taught these, and would feel liable for 'discipline' from his administrators if he taught them . . . " Beisner reported. Townley said he needed the support of the Balanced Treatment Act in order to guarantee academic freedom in the classroom.

Dr. Wayne Frair, professor of biology at The King's College in Briarcliff Manor, New York, also spoke for the defense.

"Frair testified that through his work in biochemical taxonomy (classifying forms of life by the chemicals they contain), he has become convinced that a 'limited change' model — essentially like the creation-science model — is the best explanation of living things . . . " Beisner reported. "He said that within the creation-science model, one could hold to the 'special theory' of evolution, which says that variation has taken place within certain limits among the types of life, but that one would oppose the 'general theory' of evolution, which says that all forms of life developed from a single early form and that they are therefore all genetically related."

Frair complimented the State of Arkansas by saying it is "on the very cutting edge of an educational movement" which he thought would greatly improve education because it would teach students how to think.

The defense also called Dr. Donald Chittick to the witness stand. He is a former professor of chemistry at the University of Puget Sound and currently is director of research for a company which converts biological waste materials into useable fuels.

Chittick testified that he was an evolutionist until he

entered graduate school. While there he began reading creation-science literature on the subject of origins and became convinced that it made more sense than evolution.

"Chittick pointed out that science works by beginning with certain assumptions, which it uses to interpret data, and then arrives at conclusions which are consistent with both the assumptions and the data," Beisner reported. "However, if one begins with different assumptions, he can interpret the same data and come to different conclusions."

According to Chittick, traditional methods for determining the age of the earth are unreliable. He said those methods are indicators of types of geologic forces which acted on the rocks at the time of formation and not indicative of age.

"This meant, he said, that the most highly-relied-upon method for dating the earth to be about 4.5 billion years, assumed by evolutionists, could not be used, and other methods, which indicated a young age for the earth should be used instead," Beisner reported.

Dr. N. Chandra Wickramasinghe, professor and head of the Department of Applied Mathematics and Astronomy at University College, Cardiff, Wales, was also called to testify on behalf of creation-science. Wickramasinghe is a Buddhist.

"Wickramasinghe testified that his research in partnership with Sir Federick Hoyle of the University College in astronomy and astrophysics proved beyond doubt that 'interstellar dust,' the tiny dust particles which form immense clouds in space and filter the light from some stars, are actually bits of biological material similar to the bacterium *E. coli,* a bacterium which aids in digesting food in animal colons," Beisner reported. "This discovery led Wickramasinghe and Hoyle to examine modern theories of the origin of life."

Wickramasinghe said their studies proved that the spontaneous generation of life from non-life was mathematically impossible.

"These statistical improbabilities caused Hoyle and Wickramasinghe to decide that there must be some intelligent creator either within the universe or outside it," Beisner wrote. "Wickramasinghe testified that his Buddhist background drove him to the belief that the creator was a part of the universe, but that it would be equally possible to think the creator was supernatural."

Hoyle and Wickramasinghe came up with the theory that a creator designed and formed life in the cosmos.

According to Wickramasinghe, he and Hoyle have encountered bias from the scientific community because they questioned the traditional Darwinian ideas.

"Wickramasinghe said that contrary to the popular notion that only creationism relies on the supernatural, evolutionism must as well, since the probabilities of the random formation of life (spontaneous generation) are so tiny as to require a 'miracle' for spontaneous generation to have occurred, making belief in spontaneous generation 'tantamount to a theological argument,' " Beisner wrote.

Later in an exclusive interview with Beisner, Wickramasinghe said:

• Evolutionists must imply that some kind of "scientific miracle" took place.

• Since he was a Buddhist, when he discovered scientific evidence for a creator it was quite a traumatic experience for him.

• The fact of a creator is well within the realm of empirical science.

• The denial of creation-science and a creator by many scientists is based on their anti-religious bias.

• Most scientists have a strong reaction against creation-science because of their "belief in the supremacy and centrality of man and earth . . . the human ego has been pushed right into an insignificant corner and the concept of evolution is the only way of saving that ego."

• The scientific discoveries by him and Hoyle have not been published in standard scientific journals because the editors of those journals generally won't publish anything that questions the Darwinian ideas on the origin and development of life.

The final witness for the state in defense of the Balanced Treatment Act was Dr. Robert Gentry, a research scientist at the Oak Ridge National Laboratory in Tennessee.

Gentry testified for about five hours on the subject of polonium radiohaloes. He has done extensive research on the subject.

According to Gentry, "haloes" are formed by radioactive decay of polonium in granite rocks and in wood that has turned to coal.

"The 'haloes' are circles around the decaying atoms of the radioactive material, circles which are etched into the rock by escaping alpha or beta particles that result from the decay," Beisner reported.

According to Gentry, these circles can be identified beyond question through examination under an electron microscope. He added that they are formed by all decaying radioactive materials in solidified rocks.

"Gentry said one isotope of polonium has a half-life of only about three-and-a-half minutes, meaning that in about seven minutes, the whole deposit of that isotope of polonium in a rock will have decayed away," Beisner reported from his testimony. "This means that the rock has to be solidified before or within seconds of the time the polonium isotope gets into it or there will be no halo in the rock.

"In some instances, the isotope of polonium can be given by the decay of uranium, of which the polonium isotope is one by-product. However, sometimes polonium decay haloes are found in rocks without any possible uranium source for the polonium. This, Gentry explained, means the polonium had to be present in the rocks at the moment they solidified.

"This in turn, Gentry said, means the rocks had to solidify extremely rapidly, under conditions unknown to science today. This indicates the likelihood of a creation on the earth and its . . . rocks and elements by a supernatural creator, and cannot be explained on the basis of evolutionary assumptions.

"Since evolutionary assumptions about geology insist that 'basement granites,' the granites which underlie the sedimentary layers of the earth, and the type in which Gentry often finds polonium haloes without uranium sources for the polonium, were formed by slow cooling and compaction over periods of two or three billion years, but Gentry's research shows that they must have cooled almost instantly. Gentry said his research calls into serious doubt the traditional scientific data that the earth must be 4.5 billion years old.

"Similar experiments on polonium haloes found in coalified wood led Gentry to postulate that the wood had been buried during a huge deluge and had turned rapidly to coal rather than very slowly, as evolutionary assumptions would predict. This, he said, supported the idea of the creation model that the earth's geology is best explained on the basis of a flood."

Gentry then presented letters which revealed what he called a "bias" in the scientific community against his research. He said that before the journals realized exactly what his research meant he could get articles published. But once they understood that it refuted evolution and promoted the idea of a young earth, the journals no longer accepted his articles.

Gentry noted that his research had been recognized around the world as the leading work on the subject of radiohaloes in rocks. But his research calls into question the entire field of geochronology or the assumptions about the age of the earth and its rocks.

"He said the general reaction of the evolutionist community was to discount the research, even though they could show no errors in it," Beisner reported. "He gave several examples of geologists who respond to his research simply by saying it must be wrong because if it were right it would require them to rethink all their theories about the age of the earth and the formation of the earth's geology."

Gentry said that the theories of an old earth and slow formation of the earth's rocks should be abandoned in favor of a theory of a young earth and rapid formation of the rocks.

Both the ACLU and the state made their closing arguments and left the fate of the Balanced Treatment Act in the hands of Judge Overton.

CHAPTER 10

The Decision

The die was cast.

Creationists and evolutionists had been to court. They waited impatiently for the judge to make his decision.

The question of academic freedom for school children and their indoctrination on the subject of origins rested in the hands of one man.

On January 5, 1982, Judge Overton struck down the Arkansas Balanced Treatment Act. He said, "It was simply and purely an effort to introduce the Biblical version of creation into the public school curricula."

"The argument that creation from nothing does not involve a supernatural deity has no evidentiary or rational support," he said in the ruling. "Indeed, creation of the world 'out of nothing' is the ultimate religious statement because God is the only actor."

ACLU attorney Robert Cearley, Jr., jubilant over the decision, told a news conference, "This decision has dealt the creation theory a fatal blow."

Judge Overton's 38-page decision challenged creation scientists' beliefs that the law was never intended to advance any particular religion. Rather, its intention was to provide all scientific data on the subject of origins.

The judge accused creationists of taking the book of Genesis and attempting to find scientific support for it.

"No group, no matter how large or small, may use the

organs of government, of which the public schools are the most conspicuous and influential, to force its religious beliefs on others," he said.

Judge Overton also said in the ruling:

• "Application of these two models (creation and evolution), according to creationists, and the defendants, dictates that all scientific evidence which fails to support the theory of evolution is necessarily scientific evidence in support of creationism. . . . "

• "Creation-science . . . not only fails to follow the canons defining scientific theory, it also fails to fit the more general description of 'what scientists think' and 'what scientists do.' "

• "There is . . . not one scientific journal which has published an article espousing the creation-science theory. . . . "

• "The methodology employed by creationists is another factor which is indicative that their work is not science. A scientific theory must be tentative and always subject to revision or abandonment in light of facts that are inconsistent with, or falsify, the theory. A theory that is by its own terms dogmatic, absolutist and never subject to change is not a scientific theory."

• "In efforts to establish 'evidence' in support of creation-science, the defendants relied upon the same false premise (that) . . . all evidence which criticized evolutionary theory was proof for creation. For example, the defendants established that the mathematical probability of a chemical combination resulting in life from non-life is so remote that such an occurrence is almost beyond imagination. While the statistical figures may be an impressive evidence against the theory of chance chemical combination as an explanation of origins, it requires a leap of faith to interpret those figures so as to support a complex doctrine which includes a sudden creation from nothing, a worldwide flood, separate ancestry of man and apes, and a young earth."

• "The arguments asserted by creationists are not based

upon new scientific evidence or laboratory data which has been ignored by the scientific community."

• "Robert Gentry's discovery of radioactive polonium haloes in granite and coalified wood is, perhaps, the most recent scientific work which the creationists use as argument for a 'relatively recent inception' of the earth and a 'worldwide flood.' The existence of polonium haloes in granite and coalified wood is thought to be inconsistent with radiometric dating methods based upon constant radioactive decay rates. Mr. Gentry's findings were published almost ten years ago and have been the subject of some discussion in the scientific community. The discoveries have not, however, led to the formulation of any scientific hypothesis or theory which would explain a relatively recent inception of the earth or a worldwide flood. Gentry's discovery has been treated as a minor mystery which will eventually be explained. It may deserve further investigation, but the National Science Foundation has not deemed it to be of sufficient import to support further funding."

• "Evolution is the cornerstone of modern biology, and many courses in public schools contain subject matter relating to such varied topics as the age of the earth, geology and relationships among living things. Any student who is deprived of instruction as to the prevailing scientific thought on these topics will be denied a significant part of science education."

• "Assuming for the purpose of argument, however, that evolution is a religion or religious tenet, the remedy is to stop the teaching of evolution; not establish another religion in opposition to it. Yet it is clearly established in the case law, and perhaps also in common sense, that evolution is not a religion and that teaching evolution does not violate the Establishment Clause. . . . "

• "The Court closes this opinion with a thought expressed eloquently by the great Justice Frankfurter: 'We renew our conviction that we have staked the very existence of our country on the faith that complete separation between the state and religion is best for the state and best for religion.' "

Creationists were not particularly surprised by the ruling. But they were disappointed.

The judge's decision excluding the creationists from intervention in the case sounded the death knell for the law.

There are several observations I would like to make about what went wrong in Arkansas.

1. The judge ruled that only the state's attorney general could defend the law. The attorney general and his staff attorneys undoubtedly were capable lawyers with expertise in many areas of the law. But they knew very little about the creation-evolution controversy which had simmered for generations, then boiled over in this land.

Had the attorney general endorsed the intervenors request to help in the case, the verdict could well have been quite different.

2. The attorney general and his staff failed to prepare an adequate legal defense of the law. Their discovery work and research were limited and they really didn't understand creation-science.

This basic lack of understanding manifested itself in various ways.

First, they did not know how to protect their witnesses — either during depositions or in the trial.

The ACLU strategy was to convince the judge that the Balanced Treatment Act in Arkansas was the result of a fundamentalist conspiracy in this land. Therefore, their lawyers hammered away at the religious beliefs of those who testified on behalf of the law rather than their scientific knowledge.

Second, the attorney general and his staff should have objected to the line of questioning which centered on religion rather than science. Yet the defense counsel really didn't understand the importance of restricting religious beliefs from the trial and didn't know how to do it.

Third, the attorney general should have filed motions protecting the materials subpoenaed by the ACLU from defense witnesses. Rather than seeking scientific papers, research work and proofs for creation-science from the witnesses, the ACLU majored on any religious documents the witnesses may have written or had in their possession.

The attorney general should have fought that strategy by demanding that the witnesses only be required to give in their depositions those materials which dealt with the scientific aspects of the Balanced Treatment Act.

Fourth, the attorney general did not have sufficient background in the area of creation-science to properly question the evolutionists when they were on the witness stand.

One on one, no evolutionist can debate a knowledgeable creationist and win. The creationists are the much better scientists. But that never came out in court for the attorneys for the defense never fully understood the issues or the science related to creation.

3. The attorney general should have deputized attorneys Wendell Bird and John Whitehead and turned the case over to them. Because of their vast knowledge of the issues, they could have effectively discredited the evolutionists' testimony during the trial. They would have skillfully cross-examined the evolutionists and questioned friendly witnesses in such a way as to elicit the strongest information available regarding creation.

The Arkansas Creation Science Legal Defense Fund offered cost-free legal assistance and promised to underwrite the costs for expert witnesses. But the attorney general turned it down.

A newsletter published by the group assessed the situation:

"In depositions taken by the ACLU, the attorney general's office has refused to press an objection to ACLU questions about the religious beliefs of the legislative sponsor, the model bill's draftsman, a science teacher and others in order to prevent the questions.''

According to the newsletter, the organization's attorneys urged the attorney general, during the first few depositions, to take the issue to the judge ''so that irrelevant religious issues would not prejudice the trial on balanced treatment of scientific explanations.

''The cross-examination questions by the attorney general's office in the crucial deposition of the legislative sponsor and in other depositions were 'woefully inadequate,' in the opinion of Bird and Whitehead, to counter the misleading impressions created by numerous ACLU questions. The approach of the attorney general's office has been to shoot from the hip rather than to prepare carefully weeks ahead for depositions.''

The newsletter also noted that Attorney General Clark's ''nonfeasance'' created a lot of needless headaches for some five creationist organizations. The ACLU served subpoenas on the groups — trying to find facts for the conspiracy myth — and asked for large portions of their files.

''The district court judge expressed disapproval of many of the 24 subpoena points, and instructed Clark's office to enter an agreement with the ACLU narrowing the subpoenas,'' the newsletter stated. ''The attorney general's office did not seek the protective order that it should have sought, and did not even narrow the subpoena in its agreement with the ACLU; it merely protected its own hide by getting copies of all documents turned over and limited the scope of the depositions (which was what the ACLU already had proposed).''

The result of the attorney general's non-action was that the burden was shifted to the five creationist organizations. They were forced to seek legal help to fight the subpoenas in court.

Attorney Bird assisted the Creation Science and Humanities Society and succeeded in having their entire subpoena quashed. The Institute for Creation Research also won in court and was not required to turn all their files over to the ACLU.

"The attorney general refused to assist five creationist leaders in Arkansas when the ACLU served subpoenas for virtually all of their private papers," the newsletter said. "These individuals offered to pay private attorneys to draft the necessary legal documents if the attorney general's office would merely file them and seek a protective cover. Because of the attorney general's refusal even to do that, these individuals incurred over a thousand dollars of legal expenses for work the attorney general should have done for the trial. . . .

"Bird and Whitehead have given and remain willing to give advice to the attorney general when requested, such as the names of possible expert witnesses and of useful sources . . . " the newsletter stated. "They had made arrangements to represent half of the existing defendants including the largest Arkansas school district, but those defendants were subsequently dismissed from the case."

The newsletter pointed out that attorneys Bird and Whitehead believed that Attorney General Clark's defense of the Balanced Treatment Act was "totally inadequate."

"They are 'pessimistic' about the chances for a victory for the balanced treatment of creation-science; although the legal and scientific arguments are strong they are not being adequately presented," the newsletter explained. "For this reason, several noted creation scientists have declined to serve as expert witnesses in the Arkansas trial."

The tragic result of the way the attorney general handled the pre-trial depositions was that the ACLU was allowed to force key creation scientists to focus almost entirely on religious, rather than scientific, issues.

Observers at the first week of the trial said the attorney general's staff asked very few questions on cross-examination. This prompted the judge to criticize Clark for weak cross-examination.

Those who witnessed the trial also said Clark found it

necessary to begin arguing the case by referring to God as "he or she."

"During the first week of the trial it was evident to those in the courtroom that the attorney general's staff failed to ask many obvious questions and obviously had not prepared adequately for the trial," the Legal Defense Fund newsletter said. "The attorney general made his vital objection to the admissability of personal religious beliefs without even submitting a brief or citing a case and the judge ruled against him with devastating consequences."

During the second week of the trial, things progressed from bad to worse for the creationists. The attorney general's staff continued to show a total lack of preparation.

"One witness was publicly humiliated when the judge cut him off with the exasperated comment that he had not presented one shred of scientific evidence for creation," the newsletter said. "This reflected a lack of consultation or adequate questions from the attorney general. A Dallas newspaper that editorially supported the creation cause described the attorney general's performance as the most 'bumbling' and inadequate courtroom defense that its reporters had ever seen." That newspaper was the *Dallas Morning News.*

There were many strong constitutional and scientific arguments the attorney general could have made on behalf of creation-science. One thing was certain — he either refused to make those arguments or had not made the adequate preparation necessary to do so.

M.L. Moser, Jr., a respected writer and editor in Arkansas, in a detailed account of the trial, listed several reasons why he thought the creationists lost.

"In the cross-examination it became very evident that Mr. Clark, a politician and not a skilled attorney, was completely out-classed," Moser said. "His assistants appeared to be young men fresh out of law school seeking to gain experience. Needless to say, neither Mr. Clark nor his assistants demonstrated the kind of skills and abilities required to contend with the attorneys and witnesses of the ACLU."

Moser also said Clark had publicly stated he and his staff could not devote full time to the case.

"On the other hand, during one of the breaks (in the trial) . . . I overheard a conversation between one of the attorneys brought in from New York by the ACLU. In talking with a reporter, he said that as soon as the case was filed, they put 100 attorneys to work immediately, going all over the country in search of any material that they might use in the trial."

According to Moser, another reason Clark lost the case was because he didn't use the best witnesses available for the case.

"The best witness Mr. Clark could have used was present in the courtroom every day of the trial, but was only quoted by the witnesses of the ACLU," Moser said. "Frequently the witnesses for the ACLU referred to Dr. Duane Gish, as one of the leading exponents of creation-science and they acknowledged him to be a worthy opponent for they had met him in debate, but though they made frequent references to Dr. Gish and his books and writings, Mr. Clark did not consider him an adequate witness. Dr. Gish has an earned Ph.D. degree in biochemistry from the University of California at Berkeley and has spent most of his scientific career as a research scientist with Upjohn Pharmaceutical Laboratories in Kalamazoo, Michigan, before resigning to devote full time to lecturing, debating and research in the field of creation-science. Mr. Clark did have at the end some good witnesses, nearly all from the Seventh Day Adventist Church, but by then the damage had already been done."

Moser agreed with Bird and Whitehead that the turning point in the preparation for the trial — and ultimately the trial itself — came when Judge Overton refused to allow outside lawyers and scientists to intervene and thus assist in the defense of the law.

"That meant that we would have no means of presenting any evidence or any witnesses at all unless the attorney general should choose to use them, and as earlier stated, he turned down all offers for help," Moser said. "He did not use the best witnesses available and we were at the mercy of the attorney general — and he failed us."

Moser also made an interesting point concerning the judge's decision.

"In effect, Judge Overton ruled that it was all right to teach students that men came from apes, but the public schools must say nothing about where the apes came from," he said. "It is all right to program (brainwash) school children with the creed of secular humanism (evolution), but not to expose them to the idea of a Creator, even on an equal basis."

Moser added that the result of the ruling "is that our students, our children, will go all the way through school, from kindergarten through college, and accept the theory of evolution as being the only logical explanation for the presence of the universe. . . . "

After Judge Overton made his landmark decision, attorneys Bird and Whitehead issued a brief statement on the Arkansas decision.

"The Arkansas District Court gave a constitutionally erroneous decision and factually inadequate opinion . . . " they wrote. "It is regrettable that the Arkansas defense did not adequately present or adequately support the strong constitutional arguments that could have been made in favor of balanced treatment. . . . "

The attorneys stated that teaching creation-science in the classroom in no way violates the concept of the separation of church and state.

"Creation-science can be presented solely in terms of scientific evidence and related inferences and without any religious doctrine or concepts," they said. "The primary effect and purpose are to teach all the scientific evidence on the subject of the origin of the world and life, and to restore academic freedom and free speech to public school classrooms that currently indoctrinate students in evolution-science."

The attorneys also pointed out that there were constitutional errors in the judge's ruling.

"The Arkansas court is incorrect in stating that creation-science is Genesis," they said. "Creation-science consists of scientific discussion rather than Biblical discussion, and reference to a creator or consistency with religion is permitted by the establishment clause; in fact, the Arkansas bill specifically forbids religious instruction.

"The court is also incorrect in denying that there are no scientific evidences for creation. Examples are the abrupt appearance of complex life in the fossil record and the systematic gaps between fossil types. The opinion is incorrect in its characterization of several creation-science books as essentially religious. It ignores the best public school books, misstates their content, and disagrees on scientific rather than constitutional grounds with one.

"The Arkansas court is incorrect in defining science as limited to 'natural law.' Besides the ironic religiosity of that theological term, the scientific method cannot exclude evidence on the basis of bias against the supranatural or the supernatural, and creation-science is as explanatory, testable, tentative, and falsifiable as evolution-science.

"Although the opinion recognizes that public school teachers do not have unlimited discretion under academic freedom to design their curriculum, it is incorrect in stating that public schools constitutionally may not decide not to teach a subject such as evolution-science (or calculus or art). The opinion is legally meritless in its unsupported *dictum* that instruction in evolution-science alone does not violate the free exercise rights (as well as free speech rights) of students. Numerous other constitutional errors could be cited as well."

The two expert attorneys in the field of creation-science

also said the judge's decision contained several factual inaccuracies.

Judge Overton said in his decision that "scientific creationism" did not receive recognition in America until the publication of a book entitled *The Genesis Flood* in 1965.

Bird and Whitehead pointed out that it was actually in 1974, after the publication of the book *Scientific Creationism* that the term gained credence.

The attorneys pointed out another factual error in the judge's opinion. He said that Paul Ellwanger, who had supported model legislation in various states, was motivated by a desire to see the Biblical version of creation taught in the schools.

Actually, Ellwanger opposes teaching Biblical creation in the schools, the attorneys pointed out.

Judge Overton also said that "Creation from nothing . . . and subsequent destruction of the world by flood," is unique to Genesis.

Bird and Whitehead said the two subjects are not unique to Genesis but appear in "many creation epics in human history."

The judge said that a "relatively recent inception" of the world and life means "between 6,000 and 20,000 years because of the genealogy of the Old Testament."

According to the attorneys, the recent discoveries concerning scientific data which supports a young earth has nothing to do with the Old Testament genealogy.

The judge also said in his decision that "no witness produced a scientific article for which publication had been refused" because of anti-creationist bias.

The attorneys cited Dr. Robert Gentry who in court identified several such manuscripts of his own, and various examples exist of evolutionists refusing to publish articles in scientific journals which raise questions about evolution. There are also cases where creationist students have been denied graduate degrees by the evolutionists who control the schools.

"Many other factual inaccuracies occur throughout the opinion," the attorneys said.

Perhaps we all learned a great deal by what happened in Arkansas. We learned that the ACLU and other evolutionists are well-funded, well-prepared and understand what they are doing. They are formidable opponents and the name of their game is to win.

Although we lost in Arkansas, there are other courts in other states and we dare not give up the battle for academic freedom.

CHAPTER 11

Creation-Science: True or False

Luther Sunderland, the aerospace engineer and expert on the subject of creation-science vs. evolution-science, wrote a detailed evaluation of Judge Overton's decision. Because of Sunderland's vast knowledge on the subject, let's take a look at his evaluation of the Arkansas decision so we can better understand what really happened there.

Sunderland said that when Judge Overton ruled in favor of the ACLU he did so after hearing court testimony from his own Methodist pastor and the bishop of the United Methodist Church in Arkansas.

"The entire 38-page memorandum opinion was written in a tone that clearly displayed the judge's bias which he explicitly stated from the bench in a display of temper before he heard the defendants' evidence; namely, that he thought scientific creationism should be kept in Sunday schools," Sunderland said. "If jury members openly stated such opinion about the guilt of a defendant before they heard the evidence, there is no question about their being disqualified. This is clear grounds for a mistrial, according to one network news commentator."

According to Sunderland, all the arguments presented by the ACLU can be easily refuted by expert testimony and abundance of evidence.

"If this evidence was not presented by defendants' attorneys at the time, it might have been due to their inadequate preparation and refusal to accept expert legal services offered to them," Sunderland said. "Chief defendant attorney, Arkansas Attorney General Clark, had demonstrated his lack of enthusiasm . . . by publicly stating his reservations about its constitutionality, failing to actively take depositions of witnesses during the summer, failing to object to the ACLU attorney's irrelevant religious questions during their numerous depositions, and participating in an ACLU sponsored fund-raising drive just two weeks before the trial.

"During the trial, the attorney general did such a dismal job of cross-examination that the court chastised him for not asking more questions. A Dallas newspaper reporter called this the most bumbling and inadequate defense he had ever witnessed."

Sunderland then addressed his comments to some specific points in the judge's opinion.

Judge Overton, in the decision, listed several decisions by the U.S. Supreme Court dealing with the Establishment Clause and the First Amendment. He mentioned such rulings against prayer in the schools, Bible reading and posting the Ten Commandments in the classroom.

However, Sunderland rightly pointed out that the Arkansas Balanced Treatment Act neither permitted nor advocated those rights.

Sunderland pointed out, however, that Judge Overton was "careful to avoid citing the (Supreme Court) rulings which define religion as including non-theistic beliefs such as atheism, humanism and Buddhism."

"Judge Overton first quotes a United States Supreme Court decision by Justice Black which says that government cannot 'pass laws which aid one religion, aid all religions, or prefer one religion over another,' " Sunderland said. "The state produced evidence that evolution is a basic tenet of humanism, for the first two tenets of the *Humanist Manifesto I* state, 'First: Religious humanists regard the universe as self-existing and not created. Second: Humanism believes that man is part of nature and that he emerged as the result of a continuous process.' This is clearly the definition of evolution used in most school textbooks. In the preface the *Manifesto* begins, 'Humanism is a philosophical, religious, and moral point of view as old as civilization itself.' So evolution is definitely the basis of humanism and humanism is definitely a religion."

Sunderland said Judge Overton displayed "unusual bias" by writing . . . pages of arguments as to why he believed creation-science could not be taught in the classroom without teaching the Genesis account of creation. Yet he did not mention "a single word about the state's evidence which shows that evolution is part of various religions . . . he brushes off this evidence with a simple statement, 'The argument has no legal merit.' "

The judge's opinion argued that if creation were taught alongside evolution the advancement of religion would be the primary effect, Sunderland said.

"If the state's giving students a choice of two views on origins has the primary effect of advancing religions with which creation is consistent, then why wouldn't it be far more an advancement of the religions which hold to evolution when only evolution is taught?" Sunderland asked. "The fact that evolution is normally a secular non-theistic belief has no bearing on the matter as the Supreme Court has made abundantly clear."

Sunderland said the only logical conclusion is that when schools teach only one theory, and exclude all others, they are advancing that theory.

"Teaching about two alternative theories and presenting relevant scientific evidence in an unbiased manner would only make it a fair contest," he said. "The state would not advance either theory although the evidence would obviously speak for itself if it was completely consistent with one theory and in total conflict with the other.

"The court must know that this is really the case and

that evolution could not stand up in a fair contest with creation, otherwise, why have creation scientists won all of the more than 100 debates on the subject? If evolutionists had won all of the debates, it is highly likely that the court would have welcomed the opportunity for creation to be put up alongside evolution in the classroom so it could have been discredited."

Judge Overton's opinion also gave the court's version of the controversy between creationists and evolutionists. The account is almost identical to that given by evolutionists during various debates with creationists.

"It is strange he started with the history of Christian Fundamentalism (which really began in 1910 at Princeton University with the writing of the 10 booklets called 'The Fundamentals')," Sunderland said. "The controversy actually began in earnest in 1859 when agnostics Charles Darwin and Thomas Henry Huxley attempted to eliminate the necessity of God to explain the origin of the diversity of life upon the earth. Historians of science all agree that this was Darwin's main purpose. Why then does the court blame the controversy on fundamentalists?

"Prior to Darwin, nearly all great scientists like Sir Isaac Newton, Galileo and many others, believed in creation so Darwin was the one who started the controversy. His theory was quickly accepted by atheists and agnostics but it was not until the mid-1900s that nearly all major religious denominations in this country gradually came into full acceptance of evolutionism. There has been, however, resistance to it from its popularization in 1859 and since its real beginnings with the Babylonians and Greek philosophers."

Judge Overton's preoccupation with the Christian Fundamentalists as "the major culprits in the controversy" makes one question the judge's own religious motivation, Sunderland said.

"The fact that there were more theologians than scientists in the list of plaintiffs would indicate that their motivation was religious rather than concerned about science education," he said.

According to Sunderland, the judge and most evolutionists refer to the launching of the Russian Sputnik as the turning point in the teaching of evolution. They infer that if more evolution had been taught in the schools of this country, the United States would not have been second in the space race.

"In actuality, what really happened was that the federal government did in the 1960s get careless with its money and the evolutionists capitalized on the availability of easy money using this new climate to slip evolution into school textbooks when no one was looking," Sunderland said. "They certainly never got the consent of the taxpayers. . . ."

Sunderland charged that the primary purpose of one federally funded textbook project was to entrench evolution in the school curricula and immerse primary through secondary science books in the doctrine. But Sunderland points out that Judge Overton failed to note that in 1976 the Conlan bill in Congress chastised the National Science Foundation for the biased textbooks and their treatment of origins. Congress then forced the foundation to split the textbook publishing operation off into a private, independent company.

"It should be noted that after Judge Overton gave his version of the creation-evolution controversy concentrating on the fundamentalists to show motivation . . . he then turns to Paul Ellwanger's organization which wrote the draft bill," Sunderland said. "Why does the judge fail to mention that Mr. Ellwanger is a Roman Catholic, not a fundamentalist?

"The judge included letters by Mr. Ellwanger in which he urged people to avoid religious arguments and to concentrate only on the related scientific issues," Sunderland said. "If Mr. Ellwanger had advised people to concentrate on religion because creation couldn't stand scientific scrutiny, this would have been justifiably termed deception. But like all creationists, Mr. Ellwanger was advocating fighting the battle on the basis of scientific evidence only."

"The judge would not even consider Ellwanger's testimony that he was motivated only by the denial by the public schools to open up academic inquiry and that their teaching only embraces evolution," Sunderland said.

"He gave no reason for ignoring this and chose to apply First Amendment protection to only those who embraced his own personal religious position on the matter of origins," Sunderland said. "Then he accused defendants of underhandedness in attempting to sneak their religion into schools disguised as science. . . . "

The judge also said that Ellwanger's motivation for preparing a model bill was based on his opposition to the theory of evolution, Sunderland commented, and that he wanted the Biblical theory of creation presented in the public schools.

"This is absolutely not true and no evidence was produced to prove that," Sunderland said. "It was simply a claim by the plaintiffs. There is a great difference between teaching about a theory of creation that is 'compatible' with the Biblical version and teaching the Biblical version itself."

Sunderland pointed out that the definition of creation-science as presented in the model bill "is compatible with the Biblical account but it is also compatible with other versions of creation such as the creation beliefs of all North American Indian tribes. Certainly the mention of catastrophes including a worldwide flood cannot be tied exclusively to the Bible for all world cultures have a flood tradition and there is much scientific evidence that the highest mountains were once covered with water which deposited fossils on them."

"The fact that many of the people who advocate that schools teach students about the concept of creation in an unbiased manner do believe the Bible is no justification for banning Act 590," Sunderland said. "Otherwise evolution should also be banned because it is advocated by people who believe the Bible and who claim that it is completely compatible with the Bible. Furthermore, evolution is explicitly written in the basic statement of belief of humanists — the *Humanist Manifesto*. So, using Judge Overton's arguments, evolution must also be banned from public school classrooms."

Judge Overton called the two-model approach a "contrived dualism."

"If the schools can find a third model which can be supported by scientific evidences, then that could be taught also," Sunderland said. "Creationists do say that they have never heard of a third theory which does not fit into one of the two general concepts, creation or evolution.

"Until someone comes up with a third idea, what is illogical about the creationist contention that if there is no evidence that the complexity of original life can be explained by random, purposeless processes, then it must be explained by design, purpose and pre-existing intelligence?"

Judge Overton's statement that "The scientific community does not consider origins of life a part of evolutionary theory" is one of the most preposterous statements in the entire opinion, according to Sunderland.

"The Miller and Fox experiments are presented in textbooks to show how the first cell formed spontaneously from non-living chemicals," Sunderland said. "For example, the 1980 fourth edition of *Biological Science* by Harcourt, Brace and Jovanovich . . . reads, 'Before Life Began. Astronomers, geologists and chemists are reasonably sure that the earth was formed billions of years ago, when a mass of extremely hot gases and dust particles condensed into a hot, molten (liquid) mass. . . . The primitive earth probably had an atmosphere of methane. . . . These substances must have combined gradually into increasingly complex molecules. . . . Since the molecules could self-replicate, or make copies of themselves, possibly they were a nucleic acid like DNA or RNA. Could these first replicating molecules have been a nucleic acid? Perhaps. And if they were, we would have an explanation of why DNA and RNA is the basis of inheritance in all organisms.' This book is used in schools across Arkansas, so Judge Overton based his judgment on a completely false premise. Furthermore, the state produced an expert witness, Chandra Wickramasinghe, one of the major authors of theories on the evolutionary origin of the universe. To contend that evolution theory taught in schools does not include the evolutionary origin of the universe is to claim that students are not taught about Hoyle's steady state theory or the big bang theory. How ridiculous a contention!''

The judge's ruling said that the teaching of evolution does not presuppose the absence of a creator.

"It definitely does presuppose an explanation for the origin of the universe, earth and all life without any need for God,'' Sunderland said. ''It is the explanation used and promoted by virtually all atheists.''

Overton said in his decision that the Balanced Treatment Act makes ''factually inaccurate'' statements about evolution. He cited as an example ''the sufficiency of mutation and natural selection in bringing about the existence of present living kinds from similar kinds.''

"For the last 25 or more years, the textbooks have described the mechanism of evolution in this way. It is incredible that the judge should contest this. Dr. Gould has been contesting the adequacy of natural selection and has endorsed Goldschmidt's Hopeful Monster mechanism of abrupt or rapid change,'' Sunderland said. ''This has not yet gotten into the textbooks to any significant extent although the creationists have been campaigning to have it included. For instance, the New York Education Department has been vigorously opposing inclusion of Gould's Punctuated Equilibria theory in their Biology Syllabus since 1978. The reason description of this theory is opposed so vigorously is that to admit the evidence for it, abrupt appearance and stasis, is to admit that the evidence for gradualism now taught exclusively is all wrong."

Overton was critical of the creationists using evidence which refuted evolution as evidence for creation.

"If evolution theory predicts a continuum of organisms connected with a common ancestor and creation theory predicts gaps between major different groups, why is it not a contradiction of evolution when the fossil (and living) record shows nothing but gaps?" Sunderland asked. "Since creation predicts the gaps, what possible logical reason could the court use for denying that this evidence supports creation theory? If many transitional forms were actually found connecting various groups or organisms, the court would have undoubtedly and quite properly claimed that as evidence for evolution and against creation."

According to Sunderland, the judge claims evolution starts the instant the first cell appeared on earth and then everything evolved from that common ancestor.

"His problem is that there is no testable scientific law which would have worked the DNA code of that cell up to the incredible complexity of human DNA in any amount of time, much less in 20 billion years," Sunderland observed. "So, playing by his arbitrary rules, creation theory passes the test and evolution theory flunks completely."

Sunderland said the judge "dug himself a grave" when he declared that all life started with a single cell.

"The court infers that because evolution starts with one living cell and creation theory can't define the number of original created organisms, that should make creation theory unconstitutional while evolution is OK," Sunderland said. "Where does the constitution say anything about counting ancestors?"

The judge's opinion said that the idea of "separate ancestry of man and apes" is a bold assertion.

"One could just as aptly say 'the common ancestry of man and amoeba is a bold-faced lie, explains nothing and refers to a mere religious assumption since the fossil record and scientific laws falsify it scientifically,' " Sunderland said.

Sunderland, in his evaluation of the court's decision, also said the judge's comment about the "relatively recent inception" of the earth having no scientific meaning was not an objective statement.

"There was much testimony at the trial relative to the time question which showed that it is not possible to precisely measure the age of things which existed long before recorded history because all time indicating processes used in dating require unverifiable assumptions," Sunderland said. "The 'relatively recent' term obviously has meaning in comparison with the 4.6-billion-year age for the earth mentioned in the evolution definition. The act does not mention a specific age of the earth because creation scientists have never established any precise age for the earth. They do state that students should be told about the dozens of natural processes which indicated upper limits from a few thousand years to a few hundred million years.

"It obviously frustrated the court when the act did not give a specific time, like Bishop Ussher's, to ridicule as was the case at the Scopes trial. So the court introduced Bishop Ussher's time frame so it could be used as a strawman to argue against. This demonstrates the religious motivation for the ruling."

According to Sunderland, all of the major creationist organizations agree that the question of the age of the earth should be treated as an open question and students should be presented with all the scientific data supporting the various views. However, he said that all students receive today is the data which supports the evolutionary belief that the earth is 4.6 billion years old.

"The court even admitted that Dr. Gentry's evidence for radiohaloes contradicted other radiometric dating techniques," he said.

The opinion said that not one scientific journal had pub-

lished articles supporting creation-science and added that
"it is inconceivable that such a loose knit group of independent thinkers in all the varied fields of science could, or
would, so effectively censor new scientific thought."

"It is an established fact that hundreds of active scientists with advanced degrees are creationists," Sunderland
said. "The fact that they have difficulty getting an article
published in an establishment journal demonstrates that
censorship does exist. . . . The fact that the state's
witnesses hadn't had an article rejected meant nothing, except that they were intelligent enough to not waste their
time on a useless exercise. Such action would be comparable to the United States asking Russia to broadcast an
hour of Voice of America programs on Radio Moscow.

"The court made a factual error, however, in its statement since Dr. Gentry has had several articles published in
scientific journals."

The judge also said the creation scientists failed to present any proof to support their theory.

"The court talks about proof in support of creation-
science as though there is actually such a thing as proof of
any theory in science," Sunderland said. "Its assertion that
creationists' arguments are not based upon new data has
absolutely no bearing on whether the evidence is valid or
whether it is unconstitutional for public schools to present
this evidence to students. This evidence is absolutely not
being ignored by the scientific community because prominent members such as Dr. Hoyle, Dr. Patterson and many
others are abandoning belief in evolution due to the
evidence. The complete absence of intermediate forms in
the fossil record has been well known since the time of Darwin. However, there is much new scientific data coming to
light such as information theory and computer models
which have disproved the feasibility of any mathematical
model of evolution."

Sunderland also said the court was in error when it said
that Dr. Gentry's discoveries have not led to the formation

of any new theory which would explain a young age of the earth.

"What the court means is that evolutionists have refused to consider this data like they reject all data which conflicts with established dogma. Also, there has not been a single rational explanation advanced by evolutionists showing how the polonium haloes could be explained in the 4.6-billion-year age framework for the earth."

The judge's opinion agreed with the ACLU that "it would be an infringement on teachers' academic freedom to be required to teach something with which they disagree. . . . "

Sunderland said the judge failed to "mention the rights of teachers who are forced to teach evolution although they do not consider it academically sound."

"Public polls across the country do consistently show that over 80 percent of the taxpayers want creation taught," Sunderland said. "The Constitution does not require that the state provide public education; it is done at the taxpayer's choice. As long as the taxpayers' duly elected representatives require by law that if the subject of origins is taught, the schools comply with the First Amendment and teach it objectively with no religion, how could this be illegal?"

Sunderland's evaluation of the court's decision to strike down Arkansas' Balanced Treatment Act raises some strong questions about the decision there.

What do you think?

CHAPTER 12

Creation Goes to Court

The Louisiana Legislature overwhelmingly passed Act 685 which mandated the balanced treatment of creation-science along with evolution-science. Later, Governor Dave Treen signed it, making it the law of the state of Louisiana.

However, the ink from the governor's pen was barely dry before State Superintendent of Education Kelly Nix began travelling throughout the state speaking against the new law. The tone of his speeches strongly indicated he had no intention of implementing it. That was the indication he gave during an interview with the *Baton Rouge Morning Advocate* newspaper.

On November 20, 1981, I wrote Nix a letter asking if he planned to implement the law passed by the legislature. The letter said:

"I was disturbed by an article in today's *Morning Advocate* that reported on some disagreements you and the governor are having relative to your budget requests. Two portions of that article deal with your department's preparations for the implementation next fall of the new law I authored that relates to the teaching of scientific creationism in the public schools. . . .

"The article didn't say anything about the status of preparations by your department for beginning the implementation of the law next fall, but I got the definite impression that you may be planning to wait until the beginning of the next fiscal year before commencing such preparations. I certainly hope my impression is incorrect because a July 1, 1982, start-up date for such preparations is clearly unacceptable from my viewpoint.

"If I have correctly understood your intentions, I urge you to reconsider because your department needs to utilize *all* of the time remaining between now and the start of school next fall in making necessary preparations for the implementation of this program."

Three days later I received the following reply from Superintendent Nix:

"You are fully justified in believing that I am holding up on the start of preparations for the teaching of this subject. As I see it, a great deal of work needs to be done in order to provide the parish and city school systems with the help they will need if they are going to be required to provide instruction on this rather complicated subject.

"Frankly, I find the provisions of the law rather vague. I just can't see that it makes good sense for me to spend the taxpayer's money on these extensive preparations until the courts have ruled on the constitutionality of this law.

"Therefore, I've decided that we are just going to sit tight and wait until all the legal wrangling is over before we begin the preparations for implementing this law. Further, I'm going to advise the local school superintendents not to do anything toward getting ready to teach this new subject until the courts have done their work."

Suddenly it dawned on me that we had a balanced treatment law passed through the legislative process and signed into law by the governor that might never be implemented in the classroom. Unless we filed suit to require Nix to proceed with the curriculum planning for the law.

The ACLU, during a recent news conference, had said they were preparing a suit against the Balanced Treatment Act. After thinking about the situation for several days, I decided we legislators and other interested parties should file our own suit since Superintendent Nix had informed me he had no intention of implementing the new law.

The Louisiana Legislature is filled with good and honorable men who are strong advocates of creation-science. They were deeply shocked and disappointed to learn that the law was in a state of limbo. I asked a number of them to join me in the suit and they readily accepted.

We then went to work finding others in the state — scientists, scientific organizations, parents, teachers — we thought would want to join us.

Within a short time we had a large group of people who were ready to go to court on behalf of creation-science.

Louisiana Attorney General Billy Guste agreed that the suit should be filed because the superintendent of education had refused to implement the dictates of the legislature elected by the people of the state to handle affairs of government. Guste said he would represent us and file the suit.

I also contacted attorney Wendell Bird in El Cajon, California — who is an expert on the subject of creation-science vs. evolution-science — and asked him to serve as an advisor to us in the suit. He agreed.

On December 2, 1981 — only 10 days after Superintendent Nix informed me he would not implement the law — we filed suit against him and his department in the United States District Court for the Middle District of Louisiana. We learned that Federal Judge Frank Polozola would hear the case.

Our suit asked for a declaratory judgment that the Louisiana Balanced Treatment Act is constitutional. We named as defendants the Louisiana Department of Education; J. Kelly Nix, as superintendent; the Louisiana Board of Elementary and Secondary Education; and the Orleans Parish and St. Tammany Parish School Boards which had said they would refuse to implement the law.

Attorney General Guste filed the suit on behalf of 44 legislators, scientists, public school teachers, students and parents, and religious persons of the Catholic, Jewish, Muslim and Agnostic faith.

The plaintiffs included:

• Twenty-two legislators.

• Louisiana Citizens for Academic Freedom in Origins, an organization which includes a number of scientists.

• The Committee for Openness in Science made up of some 600 scientists.

• Dr. Edward Boudreaux, internationally known chemist from New Orleans.

• Professor Stanley Morris, physics professor.

• Dr. James H.J. Hu (Ph.D.), biochemistry professor.

• Dr. John D. Waskom (Ph.D.), geology professor.

• Dr. Frank Lyon (Ph.D.), biologist.

• Professor Raymond Minchew, biology instructor.

• Jeanne C. Hart (completing Ph.D.), geologist.

• Dr. Leo T. Happel (Ph.D.), physiology professor.

• Dr. Ronald L. Williams (Ph.D.), pharmacology professor.

• Dr. James E. Rutledge (Ph.D.), and Dr. Myron H. Young (Ph.D.), both university professors.

• Dr. Mary Kleinpeter, former zoology professor.

• Dr. T.W. Christiansen, physician.

• Gloria B. Frantom, a public school life science teacher.

• Betty Buckner, a member of the Madison Parish School Board.

• Dr. Charles Harlow (Ph.D.), as a parent.

• Rev. Sam Jacobs, a Roman Catholic priest.

• Rabbinical Alliance of America, a Jewish organization.

• Rabbi Jonah Gerwitz, a Jewish rabbi.

• Dr. Asadollah Hayatdavoudi (Ph.D.), a Muslim.

• Dr. Scot Morrow (Ph.D.), an agnostic.

We caught the ACLU off guard and clearly gained momentum in the fight for balanced treatment. They filed their suit the next day in New Orleans. But we knew that from a legal standpoint our case would be the one to be heard in court.

We plaintiffs argued in our suit that the balanced teaching of creation-science with evolution-science does not violate separation of church and state because the law only requires the presentation of all of the scientific evidences on origins. We also argued that it does not violate academic freedom because it gives students a choice between the explanations.

We plaintiffs described ourselves in the suit as legislators, scientists, public school teachers, students and parents, and religious persons who believe in separation of church and state, but not in the extreme sense of separation of government from religion. We also made it clear we believe in academic freedom in public schools for students and teachers. We pointed out that we do not necessarily believe in creation-science, and in fact some are evolutionists, but we do believe in balanced treatment of the two explanations as the open-minded approach.

We emphasized that the law requires public schools to teach the scientific evidence for creation-science if they teach evidence for evolution-science, but does not require or allow instruction in any religious doctrine or material. We listed examples of what would be taught for creation-science as the abrupt appearance of complex living forms in the fossil record and the systematic gaps between fossil types, which support creation of life and particular forms.

After carefully reflecting on the Arkansas fiasco, it occurred to me that Louisiana could greatly benefit from expert legal counsel in the case. So I contacted Attorney General Guste — one of the most brilliant men in our state — and asked him to deputize attorneys Bird and John Whitehead, of Manassas, Virginia, to assist him in the defense of our law. Whitehead also is an expert in the area of creation-science and academic freedom.

The attorney general carefully examined the two attorneys' legal credentials and background and agreed that they should be used in the case. He later was so impressed with the two attorneys he asked them to be the lead counsel in the case, effectively turning the entire legal defense of the law over to them. He also committed four of his own staff attorneys to assist them.

Bird and Whitehead asked attorney Morgan Brian of New Orleans, and Tom Anderson of Palm Springs, California, to join the team defending creation-science.

We then established a legal defense fund to pay the expenses and legal fees of the attorneys. The fund was duly constituted as a non-profit corporation by the state of Louisiana.

Guste called a news conference at the capitol to announce his decision to make Bird and Whitehead lead counsel in the case.

During the news conference the two attorneys made it clear that the Louisiana law had a much better chance of standing up in court than the law which only recently had been struck down in Arkansas.

Concerning the differences between the two laws, Bird told the news conference that the law in Arkansas contained detailed definitions of evolution and creation but the Louisiana law only defines these terms as "the scientific evidences" for the two theories. He also said the Louisiana law was adopted after 12 months of legislative debate, which was not the case in Arkansas, and has the clear purpose of furthering academic freedom.

Bird also pointed out he believed the law in Louisiana would receive a better defense.

"It will be possible to present witnesses with new and different scientific evidences, with the result that the law will have a better shot in the courts," Bird said.

Guste immediately drew fire from the opponents of creation-science.

Liberal State Senator Sydney Nelson said it was inappropriate for the attorney general to file the suit on behalf of the proponents of creation-science.

Senator Nelson, who led the unsuccessful fight in the Senate to kill the bill — and only got 12 votes — apparently was still rankled because he was beaten so badly.

His sour grapes continued to show as he caustically cut down the creationists.

"It (the lobbying effort for passage of the law) was a masterful political ploy to mobilize many fundamental religious groups to support the bill," Nelson said, "and at the same time with a straight face to say it had nothing to do with religion."

The liberal state senator also criticized the attorney general by saying, "I'm disappointed our attorney general chose the maneuver he did to try to uphold the law. I think it was ill-founded and ill-conceived."

Nelson also used the well-worn, false argument that the Balanced Treatment Act is an attempt by some people to impose their morals on society.

The Shreveport Journal also editorialized against Guste and the method he was using to defend the law, namely turning the case over to Bird and Whitehead.

I discussed the *Journal* editorial and Senator Nelson's criticisms with the attorney general. We decided it was out of place for secular humanists like the editorial writer of the newspaper and the state senator to be telling us how to run our business or our court case. We both recognized they probably were upset because momentum in the case was on our side.

Now let's look at the Louisiana suit itself and examine the salient points.

Under the title "Nature of the Action" the suit declares:

"This is a First Amendment case involving academic freedom and neutral activities in the public schools. It is a dispute over the constitutionality of the Louisiana 'Balanced Treatment . . . Act. . . . ' This law was enacted to protect the academic freedom and free speech of students

and teachers in public schools by providing instruction in all
the scientific evidence about the origin of the universe and
life. . . .

"The BTA implements and enhances academic
freedom, and does not violate that First Amendment requirement, because it provides (1) that students will have
the opportunity to hear and to discuss both scientific explanations of origins if either is taught, and (2) that teachers
will have the opportunity to present a scientific explanation
that they generally do not present. Teachers will have the
responsibility to present and to discuss scientific evidence
they may personally reject if they teach a scientific explanation they may personally embrace. . . .

"The BTA fully conforms to the (First Amendment)
establishment clause, and does not constitute an establishment of religion. It requires public schools to follow a
neutral approach in teaching the scientific evidences about
the origin of the world and life, if the schools choose to present any explanations of origins. The Act specifically limits
that instruction to 'scientific evidences for (creation and
evolution) and inferences from those scientific evidences,'
without instruction in any religious doctrine. . . . "

On the subject of religious doctrine being taught in the
schools, the suit says:

"The plaintiffs agree that public school instruction in
the Biblical account of creation or religious doctrine of
evolution would violate the establishment clause, but suggest that public school instruction in the scientific
evidences for creation-science and . . . evolution-science
fully conform . . . and that the science can be taught
without the religion."

The suit also addresses the religious nature of evolution.
It says:

"Creation-science is as nonreligious as evolution-
science. . . . The concept of evolution-science is consistent
with some particular religious beliefs to the same extent
that the concept of creation-science is consistent with some

particular religious beliefs; this does not suggest that either explanation is itself inherently religious. . . . Evolution is a doctrine of a number of religious faiths (both Protestant, Catholic, Jewish, and non-Judeo-Christian) to the same extent that creation is a doctrine of a number of relogous faiths . . . but this does not preclude the existence of scientific evidence and related inferences supporting either. . . .

"Mention of the existence or nonexistence of a creator is constitutional under the establishment clause if it is neutral. . . . No court has ever held that public schools may not refer to a creator in the pledge to the flag . . . and the Declaration of Independence."

The brief also mentioned that many public school textbooks already refer to a creator or to God.

There were also several other points made in the brief:

• Academic freedom and free speech of teachers is implemented and embraced and is not violated by presenting both sides.

• It enables teachers to speak about all scientific information about origins, and to discuss alternative scientific explanations of origins, some of which have been withheld.

• The act does not forbid teachers from giving their opinions or scientific conclusions as long as both sides are presented in a balanced fashion.

• Teachers in public schools do not have an unrestricted academic freedom to teach only what they choose.

After the suit was filed, a series of pre-trial legal maneuvers began.

Attorney General Guste, Bird and Whitehead defeated the ACLU each time they met in court. Some of the victories for the creationists were impressive and proved that the ACLU could be defeated when adequate preparation was made and qualified legal assistance available.

Here are some of the victories:

• The ACLU asked that the Keith suit be thrown out of court. Our attorneys vigorously fought the motion. Their request was denied.

- The ACLU asked to intervene on behalf of the defendants, also denied.

- The ACLU asked that their suit, filed in New Orleans one day after ours was filed, be heard instead of ours. Once again our attorneys presented briefs in court opposing the motion.

Judge Adrian Duplantier in New Orleans stayed the suit thus giving our suit precedence.

- We only lost one of all the pre-trial motions. The Orleans Parish School Board asked the ACLU to join in their defense.

Attorney General Guste opposed the entrance of the ACLU into the case. He said it was incongruous for an extremist organization — which usually opposed most all state bodies and their laws — to be allowed to assist one of those bodies.

However, the ACLU was allowed to enter the case.

Because of the excellent work of the attorney general, his staff, and attorneys Bird, Whitehead, Brian and Anderson, the momentum is clearly on the side of the creationists. And we fully expect to win when we go to court — regardless of the Arkansas disaster.

CHAPTER 13

The Changing Tide — Strategy for the '80s

Sir Malcolm Muggeridge, the eminent British author, once said that future historians will look back on our generation and laugh that we believed in evolution and even taught it in our schools.

Muggeridge prophesied the demise of this unscientific evolutionary teaching and we are seeing its decline today in Louisiana, across America and around the world.

Equal time for creation laws have been enacted in Louisiana and Arkansas. The House of Representatives in Mississippi almost unanimously passed an equal time bill. The Board of Regents in New York is conducting an extensive study of creation-science in order to determine if it should be placed in the curricula of the schools of the state.

The office of the governor of Florida recently contacted my office asking for information about creation-science. State senators from California and Virginia also asked for details regarding our new law.

The Dallas, Texas, and Tampa, Florida, school systems have already implemented balanced treatment in their schools as have 13 counties in Georgia.

Bills similar to the one in Louisiana have been introduced, or are being considered, in Illinois, Iowa, Kansas, Maryland, Minnesota, Nebraska, New Hampshire, New Jersey, Ohio, Oklahoma, Oregon, Pennsylvania, South Dakota, Tennessee, Wisconsin, Washington and Colorado.

The tide is turning toward creation-science as more and more people are discovering that evolution is not science — but science fiction. When they make that great discovery they often become angry. Suddenly they realize their children have been indoctrinated in a religious dogma disguised as science. That dogma is secular humanism which focuses on man, rather than the Creator, as the center of the universe.

People are getting tired of letting evolutionists make monkeys out of them.

Parents often are quite disappointed when they learn how they have been betrayed by evolutionary teaching in the public schools. But most of the time they don't know what to do about it. So most of them just become frustrated and do nothing. They don't know how to deal with the dilemma of having creation in the home and evolution in the schools.

Also, Catholics — who could hardly be listed in the so-called "fundamentalist" camp — are speaking out on the subject of creation-science. One of their publications published an article entitled: "Why Communists Love Evolution." It said:

> The romance between Communism and evolution goes back to the days when both were in their infancy. Karl Marx found Darwin's works "the basis in natural science for our view." Thus he wrote to his friend Engels. And on another occasion he wrote: "Darwin's volume is very important and provides me with the basis in natural science for the class struggle in history." If it were not for the objection of Darwin's wife, Marx would have dedicated one volume of his *Das Capital* to Darwin.
>
> Followers of Marx in our day are no less enthusiastic over Darwinism. When Herbert Aptheker, the United States chief theoretician of Communism, lashed out against Cardinal Mindszenty during his recent visit to the United States, he accused the Cardinal and the church of having a closed mind to scientific inquiry, because Cardinal Mindszenty objected to the teaching of Darwinism when he was Primate of Hungary. Aptheker did not tell

the whole story. The Cardinal objected to the Communists forcing Darwinism in the schools of Hungary as a "discovery" of man's origin, proving the non-existence of a Supreme Being.

When the People's Army of Liberation overran his mission diocese in China, Bishop Cuthbert O'Gara relates that a mass education system was immediately set up to indoctrinate the people, not in Marxism but in Darwinism. This theory that man descended from the ape was in Bishop O'Gara's own words the "cornerstone upon which to build their new political structure." Reflecting on this in a Chinese Communism prison cell, he readily saw why the Communists began their indoctrination with evolution. "Darwinism negates God, the human soul, the after-life." In this vacuum Communism enters as the be-all and the end-all of the intellectual slavery it has created."

Much more subtle and insidious is the error that creeps into theological speculation that all things including dogma are in continual evolution. This is probably one of the main reasons why progressivist theologians and philosophers of the west are so well received in Moscow. Pope Pius XII was fully aware of their common ground back in the early fifties when he wrote the famous encyclical, *Humani Generis:*

"Some imprudently and indiscreetly hold that evolution which has not been fully proven even in the domain of natural sciences, explains the origin of all things, and audaciously support the monistic and pantheistic opinion that the world is in continual evolution. Communists gladly subscribe to this opinion so that, when the souls of men have been deprived of every idea of a personal God, they may the more efficaciously defend and propagate their dialectical materialism."

When parents, legislators or even scientists challenge the educational illusion called evolution, they soon learn how deeply entrenched it is in the public schools. They learn that it will take a wooden stake or silver bullets to strike a blow against it.

Such an experience causes them to ask:
1. Who owns the public schools?
2. Who buys the textbooks?

3. Who pays the teachers' salaries?

4. Does the taxpayer have the right to request that some concept be taught, or not taught, in the schools?

5. How did evolutionary dogma become so entrenched in the schools?

6. What can be done about it?

Attorney Wendell Bird, an authority on creation vs. evolution, relates a personal account of the great influence evolutionary teaching can have on a young mind.

Bird said he enrolled in a public school biology class in the tenth grade. The textbook said that life came about and developed through evolution and did not mention any alternative view.

His teacher assured him that evolution was the most reasonable explanation for origins and that all scientists believe it.

He did not have a background or learning in the subject of creation and greatly respected the knowledge of his teacher. He learned evolution in biology, chemistry, world history, social studies and anthropology. The result was that he came to believe in "theistic evolution" and never knew that many scientists believed another viewpoint.

Wendell Bird became a talented young scientist winning third place in the International Science Fair for a project in biology. He was also named one of the top 40 young scientists in the land by a Westinghouse Science Talent Search. But all during that time he never heard there was a scientific alternative to evolution.

Bird's experience is repeated in the lives of tens of millions of young people who enter public schools. Yet, unlike Bird, millions of them begin the school year believing in creation and then, when confronted with evolution, are never told that it is unproven.

Concerned citizens can make a difference!

If you are fed up with evolution being indoctrinated into

the minds of your children, then it's time to speak up and take a stand for truth.

Here are some things you can do:

1. Become thoroughly informed on the subject of creation vs. evolution. Read everything you can find. Vast resources are available. There are a number of scientific groups, such as the Institute for Creation Research, San Diego, California, that will provide materials to help you.

2. Once you are thoroughly informed, go to your public school science teacher and present the information to him/her asking that consideration be given to creation-science, the alternate view concerning origins. Be sure to have some good materials available for the teacher. Never be dogmatic or demanding. Make a good presentation and hope the teacher will have an open mind.

3. Many of the teachers will thank you for giving them the information and some will begin giving equal time to creation-science along with evolution-science. There is no law whatsoever which precludes teachers from teaching both theories of origins.

However, if the teacher fails to implement creation-science in the classrooms, the next step is to go to the principal and discuss it with him. Make the same presentation of factual materials and encourage him to institute it in the school. Explain to the principal that as a parent — who pays taxes to build school buildings, pays teachers' salaries, and purchases textbooks — you want both concepts presented fairly to your children.

4. Should the principal not be open to the idea, the next step is to go before the school board. Take a large group of people with you who believe like you do. Remember this when you go before any body of elected officials: you have a right to have someone represent you and your point of view. That is the basis of true democracy. So don't be hesitant. You are a taxpayer and a voter and you have every right to expect elected officials to respond to your requests.

If at all possible, it would be good to talk to the school

board members one by one prior to going before the full board. The reason for this is simple. If there is only one board member who is antagonistic toward creation-science, he might be able to sway the entire group, particularly in a public meeting. However, if each board member has had time to think about the issue and study creationist materials, he may decide it is a pretty good idea. Remember that school boards in Dallas, Texas; Tampa, Florida; Bossier City, Louisiana; and a host of other areas have ruled in favor of the balanced treatment of the subject of origins.

5. Hopefully you will get a favorable response from the board. But what if you do not? Then you need to wait until election time and elect a new board that will favor teaching both theories.

Once people realize how dangerous evolution is in the classroom it will be quite easy to rally support behind a pro-creation candidate. All the polls show that at least 75 percent of the people believe in the balanced treatment.

If you can organize only 100 people who will work for a candidate, and work hard, you can elect people to office. But it requires time, effort and a lot of hard work.

Let me explain some things you will need to do:

• Find a qualified candidate. People won't vote for just anyone, regardless of what he or she may believe.

• Raise some financial support for that candidate. He can't win a political campaign without posters, handbills, campaign cards, newspaper advertising and radio and television exposure.

• A door-to-door campaign must be carried out on behalf of the candidate by the 100 workers. Just imagine what 100 people could do working in a campaign. If 100 people visited 50 homes on a Saturday that would be 5,000 homes. During a period of one month that same 100 people could contact 20,000 homes in behalf of a candidate.

• During the door-to-door campaign on behalf of the candidate, you should make sure that you explain to each person what is happening in the schools, why you believe creation-science should be given equal time, and that your candidate is committed to the concepts of openness and fairness in the public schools.

• You should also select a candidate who is knowledge-able in various other areas of school life and has a genuine desire to help improve schools. It is wrong to support one-issue candidates and most of the general public feel strongly about those who do so.

6. Contact your senator and representative and tell them you are interested in fairness in the instruction of origins. You can write them a letter or call them on the telephone. But the very best approach would be to visit them personal-ly, discuss the subject at length and provide them with some background reading materials on the subject.

When you visit a legislator it would be good to take a half dozen other people with you so he will understand there is widespread grass-roots interest in the subject.

Public officials are elected to carry out the will of the people. It is only a myth that parents have no right to say what should and should not be presented to children in the public schools.

7. After you have contacted your legislator or other public official, then encourage your friends, relatives and neighbors to do the same.

The renowned Senator Everett Dirksen of Illinois once said that when he began to feel the heat was when he began to see the light. Most politicians feel the same way and are very sensitive to public opinion — particularly if they want to get re-elected. And believe me, that is what some of them spend the majority of their time thinking about.

Set up a speaker's bureau on the subject of creation-sci-ence and make presentations to social, civic and community clubs and other organizations. A good color slide presenta-tion would be very valuable in making the presentation.

Luther Sunderland, the great creationist aerospace engi-neer from New York, made a very impressive color slide presentation before the Senate Education Committee of the

Louisiana Legislature. It should be relatively simple, and inexpensive, to put together such a visual presentation.

8. Call on the various groups to help you in your efforts. Some of them are: The Pro Family Forum; Veterans of Foreign Wars; American Legion; large segments of the membership of free trade unions; and the majority of businessmen. Some church groups favor the balanced treatment of creation-science, others oppose it.

9. Start petition drives in your neighborhood. The most effective approach would be to divide neighborhoods up into precincts and then blocks. Choose both precinct and block captains. The block captain would take the petition to every household on the block, then report back to the precinct captain who would coordinate the efforts and compile the petition signatures.

The petitions then should be presented to the elected officials.

10. Here are some of the basics you would need to know in order to talk to elected officials about creation-science:

• There are valid scientific evidences which point to a Creator as being responsible for everything in the universe. Thousands of scientists all across America believe this.

• Creation-science is just as scientific as evolution-science.

• There are textbooks available which present the balanced treatment of the two theories.

• The law of biogenesis, universally accepted, tells us that living matter does not originate from non-living matter. The law points to creation and refutes evolution. What's wrong with presenting the law to schoolchildren?

• The Second Law of Thermodynamics says that the universe is gradually running down, not building up as the evolutionists would have us believe.

• In the fossil record, whenever man appeared he was a pure man. Monkeys were pure monkeys. All the so-called "missing links" have been proven to be hoaxes or errors. If man evolved from lower forms during a process which took millions of years, there should be billions of half-monkey half-men in the fossil record. But there are none and that points to creation.

• The creation-science movement in America today is not trying to get rid of the teaching of evolution. We are only asking that the alternative theory of creation be given equal time. This will allow the schoolchildren to make up their own minds and not be indoctrinated in only one point of view.

• You need to note that various Supreme Court rulings have said that government is supposed to remain neutral regarding religion. Yet, evolution is the religion of secular humanists, atheists, religious humanists and theological liberals. Therefore, teaching only evolution advances those religions that believe in evolution — and that's unfair.

• The balanced treatment prohibits specific religious instruction. For instance, the creation story out of Genesis would not be taught. Instruction would be limited to those scientific evidences which point to creation or evolution.

11. Remember, the school board members or legislators will never know how you feel on the subject unless you tell them.

You also need to realize that we are experiencing a counterattack from the evolutionists.

Luther Sunderland, one of the highly respected authorities on the subject of origins, says the evolutionists are preparing for all-out war against creation-science. He came to that conclusion by reading evolutionist literature published after the Arkansas trial.

According to Sunderland, this is the picture that is emerging from the evolutionist literature:

• Invite creationists to various meetings and give them a chance to make damaging statements to the creationist cause. In other words, "Give them enough rope to hang themselves."

• Try to get creationists to say they believe the Bible is

the inerrant Word of God.

Then they will claim that the creationists are not objective scientists.

Sunderland cautions creationists to "meticulously avoid discussing religion when dealing with the subject of origins in educational and scientific circles."

He also makes these additional suggestions:

• "Answer every question about religion with a reply which includes mention of the direct scientific evidence."

• "Concentrate all discussions on the only direct evidence of what actually happened in the past, relative to the origin of life — the fossil record."

• "Avoid any discussion of the interpretation of the Bible when dealing with science. Discuss only the aspects of creation theory which can be derived from observations of and deductions about scientific evidences."

• "Never discuss Bishop Ussher's chronology or six days of creation-science that can only be derived from the Bible. Stress that time is an open question and students should be permitted to hear and assess all the evidence."

• "Do not talk about a flood because there is no way a single world flood can be derived from scientific evidence alone. Talk about global catastrophes. Even the evolutionists are now saying there were about five catastrophes."

• "Don't forget, they want to trap you into talking religion (rather than science). Oblige them only if you wish to help the other side."

Remember that creation-science is pure science and has nothing to do with religion. But the strategy of the evolutionists is to try to make it appear only to be religion.

The Arkansas Democrat, in another editorial, made it quite clear that the evolutionists have a battle plan of their own. In a January 8, 1982, editorial entitled: "Evolutionist Street Fighters?" it said:

> The ACLU & Co. are naturally ecstatic over U.S. District Judge William R. Overton's upholding of their

view of Act 590 as sneaking religion. But one federal
district judge in one appeals circuit can't make constitutional law.

And since states in two other circuits (Louisiana,
Mississippi, Georgia) have either passed or are preparing to
pass differing creation-science acts of their own, the future
of creationism doesn't depend only on an Arkansas appeal
— though that appeal should certainly be made.

Meanwhile, the Arkansas trial has made a national by-
word of creationism — so much so that scientific circles are
in an uproar and talking of doing public political battle with
it — in and out of legislatures.

Maybe the Overton ruling will calm them momentarily,
but at the annual meeting of the American Association for
the Advancement of Science, scientists, science teachers
and laymen this week have been urging political battle on
establishmentarian scientists — and not just in the name of
scientific truth.

One teacher theme is separation of church and state,
which has nothing to do with science unless we believe the
Constitution requires the teaching of only scientific theory.
But would you believe a call for establishment of a fake
evolutionist orthodoxy where none now exists?

You'd perhaps expect some evolution-hipped science
teachers — who aren't, of course, scientists — to raise the
church-state issue and ask science professionals to join
them in political street battle against the creationists. But
you wouldn't expect the prestigious AAAS to sit still for the
battle plan pushed by a couple of dishonest science zealots
from Georgia — one a science teacher and the other a
member of an "educational coalition."

This prime pair told the gathering that instead of being
content with evolution as a theory and acknowledging such
doubts as afflict it, evolution-scientists should close the door
on further oral and written debates with creationists — as
well as on their own differences in theory — and use the
political forum to misrepresent evolution as immutable
truth.

Of course, some textbooks are already doing that — and
it doesn't bother judges like Overton who presume to tell us
what science really is. But we wonder that the biologists
and geologists in the AAAS hall didn't rise up and throw
these intellectual crooks out. Instead, more than one scientist there spoke up, if not for falsification, at least for doing

political battle against creationism.

One said that the issue "is not a scientific dispute, it's a political conflict." Judge Overton says it can't be that under a "constitutional system" and he may one day regret the implications of that remark, but the call to political arms was catching. An eminent Iowa scientist declared that the creationists "are working ceaselessly at grass root levels . . . using standard tactics of American politics" . . . lobbying legislators, writing letters to newspapers and distributing creationist literature. Science, he said, should do the same.

Well, it's okay for the ACLU and its anti-Christian clones and its First Amendment dupes among faculties and clergy to be anti-creationist political guerrillas. But except for court trials, science would do better to limit its activity to the adducement of scientific evidence and the writing of scholarly articles on specialties. But if science does decide to go political, the surest way for it to lose it is to falsify evolution by presenting it as truth instead of theory — which is very nearly what Overton did in his ruling.

We wouldn't suggest for a moment that they would, but if some evolution scientists do decide to fight creationism politically as non-science, we'd hope that others would define the issue not as one of science versus non-science or veiled religion but as one of free speech and academic freedom versus censorship. The right of public school teachers to teach (not preach) both evolution and creationism in a neutral "origins" context ought not even to be in dispute.

All the same, it bothers us that the scientists at the AAAS meeting could sit and listen to those moral idiots from Georgia urging them to falsify their own evolutionist data in a political offensive against creationism. That might go over with a whoop at an ACLU convention, but science professionals ought to have been outraged — and weren't.

According to the *Christian Science Monitor,* a highly respected newspaper with circulation throughout the nation, even the textbook publishers are beginning to catch on to the fact that people are fed up with evolution.

"If you look for the 'word' evolution in the index of a new high school textbook, *'Experiences in Biology,'* you won't find it," the *Monitor* said. "The reason is simple. The publisher instructed the authors to include the concept, but not the word."

The *Monitor* reported that many textbook publishers are "shying away" from evolution as the battle between creationists and evolutionists escalates.

Creationist Mel Gabler says some significant changes are being made in science textbooks regarding evolution, the *Monitor* quoted him as saying. He added that he had studied several new junior high science books and three of them had little mention of evolution.

Another sign that creationists are making a lot of progress is that they are conducting debates all over America with evolutionists and winning. An evolutionary scientist cannot face a knowledgeable creation scientist one-on-one and win. He just doesn't have the scientific facts to back up his position.

There are also some good films being produced on the creation vs. evolution controversy. One such film is: "Evolution Versus Creation: Weighing the Evidence." Quadrus Films produced the film in conjunction with the Institute for Creation Research. It is available for showing in schools, civic groups, churches and community meetings.

Evolutionists are scared because they know the creationists are the better scientists.

We have them on the run but the fight has just begun. It will take all of us working diligently together to assure that the truth of creation is taught both in our society and public schools.

There will probably always be a strong evolutionary influence in our society. And there will be men like Carl Sagan around with television programs like "Cosmos" trying to indoctrinate our children in the unscientific dogma of evolution. But Sagan's halo is about to be broken. For as the public becomes more aware of the scientific evidence for creation, they will realize how absurd evolutionist god-papa Sagan's arguments really are.

There are a host of others who, like Sagan, choose to worship ignorance. But the tide is changing.

Some Iowa State University students discovered they could make a difference in the struggle for academic freedom as it relates to creation-science. The following is a story from *Liberty,* October 1979, entitled: "Iowa State Students Protest Bias." The story is subtitled: "Academic Freedom Is a Two-Way Street — Until It Comes to the Study of Scientific Creationism." The article says:

> "You can't imagine how frustrating it is to go to a science class, to hear a professor admit the weaknesses and contradictions of evolution, to know that there is a scientific alternative, and to not even be permitted to ask questions. And if you don't think it's this drastic, you can just come to class with me anytime."
>
> Those were the words of Connie Bailey, a senior specializing in chemistry and bacteriology at Iowa State University of Science and Technology at Ames. She was one of several students who testified before the Iowa Senate in April that they had experienced academic repression in the science classrooms. Presently, the Iowa Senate is considering a bill that would require the concept of creation as supported by scientific evidence to be included along with the theory of evolution "whenever the origins of the earth or humankind is taught" in the public schools of Iowa.
>
> Scores of students rallied at the Capitol in Des Moines, before the hearing, to express their concern. The issue, they maintained, was one of academic freedom to discuss the scientific evidence for a creation and to ask questions about points of evolution in the classroom without encouraging hostility from their professors.
>
> "When I began to ask questions about some scientific evidence pointing to a creation I was met not only with indifference from my professors, but with hostility," Randy Bengfort, senior in zoology and an honor student, testified at the hearing.
>
> One student's testimony at the hearing caused him to be dismissed from a biology class when he returned to Ames.
>
> Ron Lee, senior in zoology, testified before the Senate that the scientific evidence pointing toward a creation is actually being repressed in the classroom, and those who

ask about the evidence or question evolution are being held up for public ridicule.

As a case in point, Lee told the Senate that a biology class he was taking referred to a fact about genes that the professor mentioned, and asked how the fact was consistent with the theory of evolution. "Then the professor screamed," said Lee, 'It's not a theory!' "

Lee's statement was reported in the Iowa State Daily. When Lee returned to the class, Associate Professor John Baker identified himself to the class as the professor to whom Lee had referred and, according to Lee, "said I could consider myself as being 'dropped from the course' and could leave the classroom immediately."

A letter from Lee to the senators, read before the Iowa Senate by Senator Richard Comito, emphasized Lee's dismissal as an example of "academic repression." Lee's letter explained his situation to the senators and added that whenever "the theory of scientific creationism is ever brought up in class, it is always (met) with ridicule and mocking. For instance, this past Friday (before Lee's dismissal) in my biology class, Professor Baker, in a typical mocking manner, asked if there were any creationists in the class who could tell him whether the earth was flat or round."

Lee was reinstated in the course after George Christensen, vice-president of Academic Affairs, learned of the incident. Christensen said of Lee's dismissal that it "was a thing he (Professor Baker) could not do and it should not have been done."

Baker admitted dismissing Lee from class, but publicly denied either "screaming" or saying "it is not a theory." Lee admitted later that "perhaps the word 'scream' wasn't the best word choice, but that's subjective. He did speak intently, and he did say of evolution when questioned, 'It is not a theory.' "

Students interviewed in the class did not remember the interchange between Lee and Baker, because it was at the end of class and during the students' preparations to leave. Most did remember, however, the professor asking whether any creationists could tell him whether the earth is flat or round.

Following the publicity of Lee's dismissal and reinstatement, several other students also wrote letters to the senators claiming discrimination and suppression of the scientific evidence of a creation.

> Jeff Newburn, a senior in industrial education, complained that in a biology course, "in order for me to get answers to test questions graded as correct, I had to answer as if the evolutionary theory were fact. If I answered them according to the conclusions I have drawn after studying both models, I would have received a significantly lower grade."
>
> Dan Benson, a senior in agriculture journalism, pointed out what he believed was discrimination against a popular seminar at Iowa State a few years ago that dealt with the scientific evidence for a creation. The seminar "grew quickly to 200 students," said Benson in his letter. "Most seminars average around 12, and none had ever been that large. An impromptu seminar meeting was called, and the policy was immediately changed to limit seminars to 30."

Judge Dean says there are some things parents should know about their children's rights in the public school classrooms. According to Judge Dean, there are five laws which protect children, including:

1. *Federal Hatch Amendment*

"Requires a written consent of parents in psychological testing of demeaning behavior and five or six other areas of values teaching."

2. *State Child Abuse (Physical and Mental)*

"Criminal statutes that most states have (1-10 years in Georgia if convicted). What is mental child abuse? Animal, non-theistic origins and relativistic values of deceit which are programming Humanistic Evolution Religion."

3. *"Establishment Clause" of First Amendment*

"This Constitutional right of students and parents is being violated by systematic discrimination of forcing on students humanistic, non-theistic origins and values to the exclusion of Theistic origins and absolute values."

4. *"Free Exercise Clause"*

"What is taught at home and at church is being undermined. Hostile beliefs (animal origins and relative values) are forced on students, coercion and peer pressure used. No compelling interest of state is demonstrated by state to do this. This burden may be removed by teaching both creation-science origins and absolute values. The state must be neutral — teach both or neither origin and get rid of evolving ethics games and role playing which in real life violate existing laws."

5. *"Equal Protection Clause" of Fourteenth Amendment*

"This Constitutional right is now being violated by elite educational establishments' exclusive evolving ethics and origins. Civil rights of parents and children are breached. Theists have equal protection of laws and civil rights, no less than non-theists. The law does and should not favor Humanistic Evolutionary Fundamentalists. Teachers have no legal right to force their religious views on students under the guise and pretense of teaching science and teaching students how to think. Seven Commandments of Sidney Simon are his religious absolutes forced on others."

Judge Dean has also said he believes teaching creation-science in the classroom would significantly reduce crime in America.

A biology teacher in South Dakota echoed Judge Dean's belief when he said: "I maintain that if you teach a kid that he's an animal, and that his behaviour is based on his environment, that he is going to act like an animal."

Just how important is our cause? Quite important. It is a battle for the minds of our school children. We dare not lose.

Epilogue

The fight goes on.

Creationists will soon go to court in Louisiana. There, Scopes II will enter its final phase. Hopefully we will win the great debate.

However, regardless of the outcome of the trial, the creationist movement is gaining momentum every day. People are learning what is really happening concerning the teachings of origins in our tax-supported schools — and they don't like it.

The battle for truth and academic freedom will continue forever. As long as men and women of good will burn with compassion and concern to make our land a better place to live, the fire will never die.

You, too, can join us in this great crusade for freedom of speech, freedom to know the truth and freedom from educational oppression and indoctrination.

Creation-science is pure science and it belongs in the public school classrooms. Yet censors abridge it from the curricula. It's time the people of America insist that the wrong be corrected.

It can be done in our time.

MURDERED HEIRESS ... LIVING WITNESS, by Dr. Petti Wagner, $5.95.

This is the book of the year about Dr. Petti Wagner — heiress to a large fortune — who was kidnapped and murdered for her wealth, yet through a miracle of God lives today.

Dr. Wagner did indeed endure a horrible death experience, but through God's mercy, she had her life given back to her to serve Jesus and help suffering humanity.

Some of the events recorded in the book are terrifying. But the purpose is not to detail a violent murder conspiracy but to magnify the glorious intervention of God.

THE HIDDEN DANGERS OF THE RAINBOW: The New Age Movement and Our Coming Age of Barbarism, by Constance Cumbey, $5.95.

A national best-seller, this book exposes the New Age Movement which is made up of tens of thousands of organizations throughout the world. The movement's goal is to set up a one-world order under the leadership of a false christ.

Mrs. Cumbey is a trial lawyer from Detroit, Michigan, and has spent years exposing the New Age Movement and the false christ.

Feel Better and Live Longer Through **THE DIVINE CONNECTION,** by Dr. Donald Whitaker.

This is a Christian's guide to life extension. Dr. Whitaker of Longview, Texas, says you really can feel better and live longer by following Biblical principles set forth in the word of God.

THE DIVINE CONNECTION shows you how to experience divine health, a happier life, relief from stress, a better appearance, a healthier outlook, a zest for living and a sound emotional life. And much, much more.

THE AGONY OF DECEPTION, by Ron Rigsbee with Dorothy Bakker, $6.95.

Ron Rigsbee was a man who through surgery became a woman and now through the grace of God is a man again. This book — written very tastefully — is the story of God's wondrous grace and His miraculous deliverance of a disoriented young man. It offers hope for millions of others trapped in the agony of deception.

— What happened in the Arkansas Creation trial?

— Why did the creationists lose?

— Where was the Christian media during the trial?

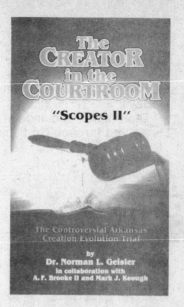

The secular media had a field day in the 1925 "evolution" trial. In the 1981 Arkansas trial, things weren't much better. Sadly, not one single journalist from a Christian paper showed up to cover "Scopes II". Because of this, even the coverage in Christian publications was influenced by the distorted reports in the secular press.

The true record must become known to every earnest Christian. To build for triumph in the courts in the future, we must understand our past failures. The issues are too important to ignore. As Norm Geisler writes in *The Creator*:

"What happened in Arkansas should arouse all freedom-loving people."

We share Dr. Geisler's concern.

MOTT MEDIA

1000 E. Huron Street
Milford, MI 48042

Available at fine
Christian Bookstores
everywhere.

Yes, send me the following books:

_____ copy (copies) of **Murdered Heiress . . . Living Witness** @ $5.95 _____ =

_____ copy (copies) of **The Hidden Dangers of the Rainbow** @ $5.95 _____ =

_____ copy (copies) of **The Divine Connection** @ $4.95 _____ =

_____ copy (copies) of **The Agony of Deception** @ $6.95 _____ =

_____ copy (copies) of **Training for Triumph** @ $4.95 _____ =

_____ copy (copies) of **The Day They Padlocked the Church** @ $3.50 _____ =

_____ copy (copies) of **Backward Masking Unmasked** @ $4.95 _____ =

_____ copy (copies) of **Scopes II/The Great Debate** @ $4.95 _____ =

_____ copy (copies) of **Why J.R.?** @ $4.95 _____ =

_____ copy (copies) of **Need a Miracle?** @ $4.95 _____ =

_____ copy (copies) of **Yes, Lord!** @ $4.95 _____ =

Enclosed is: $ _____ including postage *(please enclose $1 per book for postage)* for books.

NAME _____

ADDRESS _____

CITY AND STATE _____ ZIP _____

Mail to HUNTINGTON HOUSE, INC., P.O. Box 78205, Shreveport, Louisiana 71137
Telephone Orders: (TOLL FREE) 1-800-572-8213, or in Louisiana (318) 222-1350